Descriptive Statistics

A Contemporary Approach

Richard P. Runyon
C. W. Post College,
Long Island University

Descriptive Statistics

A Contemporary Approach

Addison-Wesley Publishing Company

Reading, Massachusetts
Menlo Park, California
London · Amsterdam
Don Mills, Ontario · Sydney

This book is in the
Addison-Wesley Series in Statistics

ISBN 0-201-06652-1
ABCDEFGHIJ-AL-7987

This book covers beginning descriptive statistics. It has been designed to accomplish in the field of Statistics what Professors Mervin L. Keedy and Marvin L. Bittinger have achieved in the field of mathematics with their unique series devoted to modern instruction in arithmetic and various levels of algebra.

The book bears the words "A Contemporary Approach" in its title to emphasize two important characteristics: (1) the relevance of statistics to application in a broad spectrum of contemporary disciplines; and (2) the possibility of achieving conceptual and computational mastery of the subject matter without using a deadening array of complicated formulas and esoteric mathematical proofs.

Descriptive Statistics represents many years of cumulative experience in both classroom instruction and textbook construction in the field of statistics. It represents an effort to apply my comprehension of the learning process to a field which many students approach with both awe and apprehension. Several noteworthy features in this respect are:

1. The format permits a student to follow the step-by-step solution to a given statistical problem and then apply the knowledge immediately to a series of exercises that appear in the margins of the text. We know that the immediacy of feedback enhances the learning and retention processes.

2. The student is then able to check the accuracy of an answer by referring to the worked solutions that appear at the end of each chapter. In fact, where called for, the solutions are detailed so that the student will be able to locate the source of error whenever an answer disagrees with the one appearing in the text. There are over five hundred of these exercises and answers.

3. Review appears repeatedly throughout the book. For example, students are first introduced to the concept and calculation of a percentile rank in Chapter 4. However, the calculation of the percentile rank is essential in transforming a nonnormal distribution of scores into a normal distribution (Chapter 8). The procedures for calculating a percentile rank are repeated at this point.

4. Because of this built-in review feature, individual chapters can stand by themselves. This feature makes the book valuable outside the classroom as a reference on computational methods. For example, if a researcher wants to have an assistant express a set of scores in units of the standard normal curve, he or she may direct the assistant's attention to Chapter 8.

5. The table of Percent of Area Under the Standard Normal Curve has been completely revised to eliminate common sources of student difficulty. To begin with, both positive and negative values of z are shown. As a result, it is no longer

necessary to give detailed and confusing instructions on how to find, for example, the percentile rank of negative z as opposed to a positive z. Percentile ranks are given directly. The new arrangement also simplifies calculations that require finding areas between various z-scores.

There are a variety of ways in which the book can be used. It can serve by itself as a first course in statistics. Although computational procedures are emphasized, there is a great deal of conceptual material introduced into illustrative examples and exercises. Because of the "building-block" nature of the book's development, it is ideally suited for the self-paced courses appearing with increasing frequency on the academic scene. I am referring to courses variously labelled as Personalized Instruction, PSI, and the like. The Test Booklet will be particularly useful in such courses. Each chapter includes three ditto masters of alternate forms of text instruments.

Descriptive Statistics: A Contemporary Approach may also be used as a supplementary text in courses for which a regular text has been assigned. Altogether too often, in my experience, students will falter because their inability to calculate a given statistic will produce negative side effects that interfere with conceptual processes. The instructor who uses this as a supplementary text will help the students avoid this pitfall. He or she may then feel free to devote class time to the important cognitive aspects of statistical analysis. This text integrates well with most books on the market since it covers most topics with a minimum of computational formulas and provides alternate sections in areas where there is a lack of uniformity in current procedures. For example, some instructors prefer to use a single defining formula for the standard deviation in both descriptive and inferential statistics. They always use $N-1$ in the denominator of the standard deviation. Other instructors prefer to define the standard deviation by N in the denominator when handling descriptive statistics. This book includes two sets of parallel instructions and exercises, differentiated from one another by the use of N or $N-1$ in the denominator of the standard deviation.

Acknowledgements

I wish to thank my son Richard O. Runyon for preparing the new table of Percent of Area Under the Standard Normal Curve. I wish also to express my deepest appreciation to Nancie Brownley for her care, perseverance, and good humor while converting my chicken scratches into a legible manuscript.

November 1976   R.P.R.

Contents

Review of Basic Mathematics

This first chapter contains a review of the basic mathematical operations necessary to complete the course in statistics. You should complete this chapter in order to avoid stumbling in this course as a result of inadequate recollection of arithmetic and algebraic operations.

ADDITION AND SUBTRACTION

When you add a series of numbers, the order of adding the numbers has no influence on the sum.

Examples

$8 + 5 + 3 = 3 + 5 + 8 = 8 + 3 + 5 = 5 + 8 + 3$

$4 + 0 + 6 + 5 = 0 + 4 + 5 + 6 = 5 + 6 + 0 + 4$

Do Exercise 1 at the right.

If you add a series of positive and negative numbers, the order of adding the numbers has no influence on the sum. However, it is often convenient to group together the numbers preceded by a positive sign, to group together the numbers preceded by a negative sign, to obtain each sum separately, and to subtract the latter sum from the former.

Example

$2 + 6 - 5 + 3 - 4 + 6 = 2 + 6 + 3 + 6 + (-5 - 4)$

$= 17 - 9$

$= 8$

To subtract a larger numerical value from a smaller numerical value, we ignore the signs, subtract the smaller number from the larger, and affix a negative sign to the sum.

Example

$5 - 6 + 4 - 9 + 3 - 2 + 1 = 5 + 4 + 3 + 1 - 6 - 9 - 2$

$= 13 - 17$

$= -4$

Do Exercises 2, 3, and 4 at the right.

MULTIPLICATION

The order in which numbers are multiplied has no effect on the product.

Examples

$4 \times 6 \times 7 = 7 \times 4 \times 6 = 4 \times 7 \times 6 = 168$

$5 \times 12 \times 2 = 2 \times 5 \times 12 = 120$

Do Exercises 5 and 6 at the right.

If you have a mixed expression involving multiplication, subtraction, and addition of single terms, multiplication is performed first.

Examples

$$6 \times 5 - 4 + 3 \times 2 = 6 \times 5 + 3 \times 2 - 4$$
$$= 30 + 6 - 4$$
$$= 32$$

$$-8 \times 6 + 5 \times 12 - 30 = -48 + 60 - 30$$
$$= -78 + 60$$
$$= -18$$

Do Exercises 7 and 8 at the right.

If the problem involves finding the product of one term multiplied by a second expression, which includes two or more terms which are either added or subtracted, you may multiply first and then add (or subtract) or add (or subtract) first and then multiply.

Examples

$$4(6 - 3) = 4 \times 6 - 4 \times 3$$
$$= 24 - 12 = 12$$

or

$$4(6 - 3) = 4(3)$$
$$= 12$$

$$9(8 - 12) = 9 \times 8 - 9 \times 12$$
$$= 72 - 108$$
$$= -36$$

or

$$9(8 - 12) = 9(-4)$$
$$= -36$$

Multiply.

5. If $6 \times 5 \times 7 = 210$,

   $5 \times 7 \times 6 = $ _____ .

6. If $8 \times 0.4 \times 0.2 = 0.64$,

   $0.2 \times 8 \times 0.4 = $ _____ .

7. $15 \times -8 + 16 - 5 \times 7 = $ _____

8. $2 \times 0 + 15 - 9 \times 3 \times -2$

   $= $ _____

In most cases, it is more convenient to reduce the expression within the parentheses first.

Do Exercises 9, 10, and 11 at the right.

If two numbers having like signs are multiplied together, the product is always positive.

Examples

$6 \times 5 = 30$ $-4 \times -9 = 36$

Do Exercises 12 and 13 at the right.

If two numbers having unlike signs are multiplied together, the product is always negative.

Examples

$-8 \times 15 = -120$ $11 \times -3 = -33$

Do Exercises 14 and 15 at the right.

The same rules apply to division.

Examples

$$\frac{-10}{-5} = 2 \qquad\qquad \frac{-12}{3} = -4$$

Do Exercises 16 and 17 at the right.

Multiplication is a special case of successive addition.

Examples

$3 \times 2 = 2 + 2 + 2 = 3 + 3$

$9 \times 7 = 7 + 7 + 7 + 7 + 7 + 7 + 7 + 7 + 7$

$\qquad = 9 + 9 + 9 + 9 + 9 + 9 + 9$

$15 + 15 + 15 + 12 + 12 + 9 + 9 + 9 + 9$

$\qquad\qquad\qquad = 3(15) + 2(12) + 4(9)$

$\qquad\qquad\qquad = 45 + 24 + 36$

$\qquad\qquad\qquad = 105$

Do Exercises 18 and 19 at the right.

9. $7(16 - 5) =$ _____

10. $8(14 - 16) =$ _____

11. $8(12 - 15) + 6(4 - 2) =$ _____

12. $8 \times 7 =$ _____

13. $-8 \times -9 =$ _____

14. $6 \times -5 =$ _____

15. $8(4 - 6) \times 7 =$ _____

16. $\dfrac{9(3 - 5)}{3(7 - 9)} =$ _____

17. $\dfrac{6(7 - 12)}{6(10 - 5)} =$ _____

18. $1 + 1 + 1 + 1 + 6 + 6 + 6 + 6 + 6$
    $+ 7 + 7 + 8 + 8 + 8 =$ _____

19. $2 + 2 + 2 + 2 - 3 - 3 - 3 - 3 - 3$
    $+ 4 + 4 + 4 =$ _____

ALGEBRAIC OPERATIONS

Transposing

To transpose a term from one side of an equation to the other side, you merely change the sign of the transposed term.

Examples

We see that

$x + y = z$

$x = z - y$

$0 = z - y - x$

are all equivalent statements.

So also are these:

$3 + 2 = 5$

$3 = 5 - 2$

$0 = 5 - 2 - 3$

Do Exercises 20 and 21 at the right.

Solving Equations Involving Fractions

The important principle to remember is that *equals multiplied by equals are equal.*

Examples

Solve the following equation for x.

$b = a/x$

Multiply both sides of the equation by x/b.

$$\frac{\cancel{b}x}{\cancel{b}} = \frac{a\cancel{x}}{\cancel{x}b}$$

This reduces to

$$x = \frac{a}{b}.$$

Solve the equation

$$b = \frac{a}{x} \qquad \text{for } a.$$

20. Solve for a.

$a + b + c = 19$

21. Solve for y.

$20 + n = y + 2$

Multiply both sides by x.

$$b \cdot x = \frac{a}{\not x} \cdot \not x$$

This reduces to $bx = a$ or $a = bx$.

Do Exercises 22 through 24 at the right.

Two simple rules can be employed to transpose terms:

1) A term which is in the denominator on one side of the equation may be moved to the other side of the equation by multiplying it by the numerator on that side.

Example

Solve $\frac{x}{a} = b$ for x.

$x = ab$

2) A term in the numerator on one side of the equation may be moved to the other side of the equation by dividing into the numerator on that side.

Example

This means that $ab = x$ may become

$$a = \frac{x}{b} \qquad \text{or} \qquad b = \frac{x}{a}.$$

Do Exercises 25 through 28 at the right.

MULTIPLICATION AND DIVISION OF TERMS HAVING EXPONENTS

An exponent tells how many times a number is to be multiplied by itself.

Examples

$$X^5 = X \cdot X \cdot X \cdot X \cdot X \qquad 2^6 = 2 \cdot 2 \cdot 2 \cdot 2 \cdot 2 \cdot 2$$

Multiplication Rule

To multiply X raised to the ath power (X^a) by X raised to the bth power (X^b), you simply add exponents and raise X to the $(a + b)$th power.

22. Solve the following for ΣX.

$$\bar{X} = \frac{\Sigma X}{N}$$

23. Solve the above for N.

24. Solve for $n - 1$.

$$s^2 = \frac{\Sigma x^2}{n - 1}$$

25. Solve for ΣX.

$$\bar{X} = \frac{\Sigma X}{N}$$

26. Solve the above for N.

27. Solve for Σx^2.

$$s^2 = \frac{\Sigma x^2}{n - 1}$$

28. Solve the equation in Exercise 27 for $n - 1$.

Examples

Given: $X^a \cdot X^b = X^{a+b}$.

If $a = 3$ and $b = 5$,

$X^{a+b} = X^8 = X \cdot X \cdot X \cdot X \cdot X \cdot X \cdot X \cdot X$.

$(1/2)^3 \cdot (1/2)^4 = (1/2)^7$

$\qquad\qquad = 1/2 \cdot 1/2 \cdot 1/2 \cdot 1/2 \cdot 1/2 \cdot 1/2 \cdot 1/2$

$\qquad\qquad = 1/128$

Do Exercises 29 through 31 at the right.

Division Rule

To divide X raised to the ath power by X raised to the bth power, you simply subtract the exponents in the denominator from the exponents in the numerator.

Examples

$$\frac{X^n}{X^a} = X^{n-a} \qquad\qquad \frac{5^4}{5^1} = 5^3 = 125 \qquad\qquad \frac{2^5}{2^7} = 2^{-2}$$

Note that the last example above may be stated

2^{-2} or $(1/2)^2$.

Thus, $2^{-2} = (1/2)^2 = 1/4$, and $4^{-4} = (1/4)^4 = 1/256$.

Do Exercises 32 through 36 at the right.

Extracting Square Roots

The square root of a number is the value which, when multiplied by itself, equals that number.

Most square roots may be obtained directly from existing tables. The usual difficulty is determining how many digits precede the decimal. For example, $\sqrt{225,000,000} = 5,000$; not 500 or 50,000. There are four digits before the decimal.

To calculate the number of digits preceding the decimal, count the number of pairs to the left of the decimal:

number of pairs = number of digits.

However, if there is an odd number of digits, the number of digits preceding the decimal equals the number of pairs + 1.

29. If $a = 4$ and $b = 6$, find X^{a+b}.

30. Multiply 6^5 times 6^3.

31. Multiply $(1/4)^2$ times $(1/4)^4$.

32. $\dfrac{X^y}{X^a} =$ _____

33. $\dfrac{4^6}{4^3} =$ _____

34. $\dfrac{2^2}{2^5} =$ _____

35. $\dfrac{7^4}{7^7} =$ _____

36. $\dfrac{3^5}{3^5} =$ _____

Examples

a) $\sqrt{\overset{50.0}{2500.00}}$ b) $\sqrt{\overset{15.8}{250.00}}$

c) $\sqrt{\overset{5.0}{25.00}}$ d) $\sqrt{\overset{1.58}{2.5000}}$

Do Exercises 37 through 42 at the right.

ROUNDING

There are two types of situations in which we resort to rounding:

1) When we divide one whole number into a second and obtain either a long or infinite remainder (for example, $1/3 = 0.333333+$), and
2) when we measure variables, such as height, weight, or temperature, that are never measured exactly.

The rule is simple: Decide the number of places to which you want to express the final answer. Suppose you want the answer to be rounded to the second decimal place. Examine the entire remainder beyond the second decimal place. If it is *greater* than 5, raise the number at the second decimal place to the next higher number.

Examples

68.57602 = 68.58   0.06943 = 0.07

8.495001 = 8.50

If it is *less* than 5, allow the digit at the second place to remain as is.

Examples

67.50499 = 67.50   0.08099 = 0.08

6.49362 = 6.49

If it is *exactly* 5, with no remainder, round the digit at the second decimal place to the nearest even number.

Examples

68.49500 = 68.50   0.08500 = 0.08

68.50500 = 68.50

Do Exercises 43 through 52 at the right.

For Exercises 37 through 42 refer to Table A-3 in the Appendix.

37. Find $\sqrt{30.00}$.

38. Find $\sqrt{300.00}$.

39. Find $\sqrt{3000.00}$.

40. Find $\sqrt{30000.00}$.

41. Find $\sqrt{169.00}$.

42. Find $\sqrt{1600.00}$.

43. Round to two decimal places.
    0.05489

44. Round to one decimal place.
    78.4339

45. Round to three decimal places.
    85.46749

46. Round to two decimal places.
    0.05501

47. Round to one decimal place.
    0.05501

48. Round to the unit's digit.
    648.73

49. Round to the second decimal place.
    69.99501

50. Round to the first decimal place.
    70.7500

51. Round to the first decimal place.
    70.6500

52. Round to the third decimal place.
    70.650500

CHAPTER 1 TEST

1. Find the sum of the following numbers: $12 + 5 - 3 - 4 + 2$
   a) 26   b) 8
   c) 10   d) none of the preceding

2. Find the sum of the following numbers: $-15 + 5 - 8 - 3 + 12$
   a) -9 b) -11 c) 19   d) 21

3. The product of $6 \times 4 \times 7$ is the same as:
   a) $5 \times 5 \times 7$ b) $12 \times 2 \times 3.5$
   c) $4 \times 7 \times 6$ d) none of the preceding

4. $16 - 8 \times 4 + 3$ equals:
   a) -13 b) 56   c) 29   d) 35

5. $12(7 - 4)$ equals:
   a) 80   b) 36   c) -36 d) 88

6. $7(8 - 14)$ equals:
   a) 42   b) -42 c) 1   d) -90

7. $-5 \times 6 \times 3$ equals:
   a) 90   b) -33 c) 13   d) -90

8. $-6 \times -4 \times 3$ equals:
   a) 72   b) -30 c) -18 d) -72

9. $\dfrac{10 - 15}{-5}$ equals:

   a) -1 b) $3/2$ c) $-3/2$ d) 1

10. Another notation for $6 + 6 + 6 + 6 + 6 + 6$ is:
    a) 6^6 b) $6 + 6$ c) 6×6 d) 3×6^2

11. Another notation for $5 + 4 + 4 + 4 + 4 + 4 + 3 + 3 + 3 + 2$ is:
    a) $5^1 + 4^5 + 3^3 + 2^1$
    b) $5 + 4 \times 5 + 3 \times 2 + 2$
    c) $5 + 4 + 3 + 2$
    d) $1 + 5 + 3 + 1(5 + 4 + 3 + 2)$

12. $s^2 = \dfrac{\Sigma x^2}{N - 1}$ is the same as:

    a) $N - 1 = \dfrac{s^2}{\Sigma x^2}$ b) $s^2 - (N - 1) = \Sigma x^2$

    c) $s^2 - \Sigma x^2 = N - 1$ d) $N - 1 = \dfrac{\Sigma x^2}{s^2}$

13. Select the expression that is *not* equivalent to the remaining three:

a) $r = \dfrac{\Sigma xy}{\sqrt{(\Sigma x^2)(\Sigma y^2)}}$ b) $\sqrt{(\Sigma x^2)(\Sigma y^2)} = \dfrac{\Sigma xy}{r}$

c) $r - \dfrac{\Sigma xy}{\sqrt{(\Sigma x^2)(\Sigma y^2)}} = 0$ d) $\sqrt{(\Sigma x^2)(\Sigma y^2)} = \dfrac{r}{\Sigma xy}$

14. Select the expression that is *not* equivalent to the remaining three:

a) $l + m - 15 + x = y$ b) $l + m + x = y + 15$

c) $l - y + m - 15 + x = 0$ d) $l - 15 = y - (m - x)$

15. Solve the following equation for N: $\bar{X} = \dfrac{\Sigma X}{N}$

a) $N = \dfrac{\Sigma X}{\bar{X}}$ b) $N = \dfrac{\bar{X}}{\Sigma X}$

c) $N = \Sigma X - \bar{X}$ d) $N = \bar{X} - \Sigma X$

16. Solve the following for f: $g + \dfrac{f}{c} - 8 = 0$

a) $f = 8 - g - c$ b) $f = \dfrac{8 - g}{c}$

c) $f = c(g - 8)$ d) $f = c(8 + g)$

17. $(P^3)(P^6)$ equals:

a) P^{18} b) P^{-3} c) P^3 d) P^9

18. If $P = 3$, what is the value of P^{-3}?

a) -9 b) -27 c) $1/27$ d) 1

19. $\dfrac{P^{12}}{P^6}$ equals:

a) P^6 b) P^2

c) P^{18} d) none of the preceding

20. $\dfrac{P^{15}}{P^{20}}$ equals:

a) P^{-5} b) $P^{3/4}$ c) P^{35} d) P^{300}

21. If $P = 5$, $\dfrac{P^8}{P^4}$ equals:

a) 2 b) 10 c) 625 d) 20

22. If $P = 2$, $\dfrac{P^4}{P^8}$ equals:

a) $1/2$ b) 16 c) $1/16$ d) 1.00

For Problems 23 through 25, refer to Table A-3 in the Appendix.

23. $\sqrt{1500}$ equals:

   a) 38.730   b) 3.8730   c) 12.2474   d) 122.474

24. $\sqrt{9.30}$ equals:

   a) 9.6437   b) 30.4959   c) 0.96437   d) 3.04959

25. $\sqrt{0.015}$ equals:

   a) 0.38730   b) 12.2474   c) 0.122474   d) 0.038730

26. Rounded to two decimal places, the number 0.55500 is:

   a) 1.00   b) 0.56   c) 0.60   d) 0.55

27. Rounded to the nearest whole number, the number 0.5001 is:

   a) 0   b) 1   c) 0.50   d) 1.5

28. Rounded to the third decimal place, the number 84.659501 is:

   a) 84.659   b) 84.660   c) 84.700   d) 85.000

29. Rounded to the third decimal place, the number 9.9952 is:

   a) 9.995   b) 9.100   c) 10.000   d) 9.996

30. Rounded to the second decimal place, the number 0.73499 is:

   a) 0.74   b) 1.00   c) 0.70   d) 0.73

Basic Mathematical Concepts

SUMMATION NOTATION

The Greek capital letter sigma, Σ, is commonly employed to direct us to sum the quantities following it. Thus, if I wanted you to sum the quantities

$$X_1 = 12, \quad X_2 = 15, \quad X_3 = 35, \quad X_4 = 12,$$

you could write

$$\Sigma(X_1, X_2, X_3, X_4).$$

Examples

Sum the quantities X, Y and Z.

$$\Sigma(X, Y, Z)$$

Of five quantities, X_1, X_2, X_3, X_4, X_5, sum X_3, X_4, X_5.

$$\Sigma(X_3, X_4, X_5)$$

Do Exercises 1, 2, and 3 at the right.

These notations may be abbreviated by the use of subscripts and supercripts. To indicate the summation

$$\Sigma(X_1, X_2, X_3, X_N),$$

we can write

$$\sum_{i=1}^{N} X_i.$$

This directs us to sum all quantities starting with i and proceeding through N. The subscript i indicates the value of X at which the summation is to begin.

In the above example, $i = 1$ means we should start summing with the quantity X_1.

Examples

Symbolic notation for summing quantities X_2 through X_4:

$$\sum_{i=2}^{4} X_i$$

Symbolic notation for summing quantities X_2 through the final quantity (X_N):

$$\sum_{i=2}^{N} X_i$$

Do Exercises 4 through 7 at the right.

EXERCISES

Write scientific notation for summing the following quantities.

1. X_5, X_7, X_{10}

2. 3, 7, 12, 15

3. A, C, Y

Given the quantities

$$X_1, X_2, X_3, X_4, X_5, X_6, X_7, X_N,$$

write summation notation for the following.

4. Adding quantities X_5 through X_N

5. Adding quantities X_1 through X_6

6. Adding quantities X_1 through X_N

7. Adding quantities X_4 through X_7

2

Now assume the values of each of the quantities or variables to be

$$X_1 = 5, \quad X_2 = 3, \quad X_3 = 8, \quad X_4 = 0.$$

We can use symbolic notations to find the sum of different values of the variable X.

Examples

Find $\sum\limits_{i=1}^{N} X_i$.

$$\sum\limits_{i=1}^{N} X_i = 5 + 3 + 8 + 0$$
$$= 16$$

Find $\sum\limits_{i=3}^{N} X_i$.

$$\sum\limits_{i=3}^{N} X_i = 8 + 0$$
$$= 8$$

Find $\sum\limits_{i=2}^{N} X_i$.

$$\sum\limits_{i=2}^{N} X_i = 3 + 8 + 0$$
$$= 11$$

Do Exercises 8 through 12 at the right.

SUMMATION RULES

The sum of the values of a variable with a constant added to each value is equal to the sum of the values plus N times that constant. Stated symbolically:

$$\sum\limits_{i=1}^{N} (X_i + a) = \sum\limits_{i=1}^{N} X_i + Na.$$

Examples

Given the following values:

$$X_i = 4, \quad X_2 = 6, \quad X_3 = 5.$$

Add a constant of $a = 2$ to each value.

$$\sum\limits_{i=1}^{3} (X_i + a) = 15 + 3(2)$$
$$= 21$$

Add a constant $a = 5$ to each value.

$$\sum\limits_{i=1}^{3} (X_i + a) = 15 + 3(5)$$
$$= 30$$

Do Exercises 13 through 15 at the right.

Given the following values of the variable X:

$$X_1 = 4, \ X_2 = 7, \ X_3 = 2, \ X_4 = 0,$$
$$X_5 = 9, \ X_6 = 10, \ X_7 = 1, \ X_8 = 3.$$

find the indicated sums.

8. $\sum\limits_{i=1}^{N} X_i$

9. $\sum\limits_{i=4}^{7} X_i$

10. $\sum\limits_{i=2}^{N} X_i$

11. $\sum\limits_{i=6}^{7} X_i$

12. $\sum\limits_{i=1}^{5} X_i$

Given the following values of the variable X:

$$X_1 = 4, \quad X_2 = 7, \quad X_3 = 2, \quad X_4 = 0,$$
$$X_5 = 9, \quad X_6 = 10, \quad X_7 = 1, \quad X_8 = 3.$$

13. Find $\sum\limits_{i=1}^{N} (X_i + a)$, when $a = 1$.

14. Find $\sum\limits_{i=1}^{N} (X_i + a)$, when $a = 5$.

15. Find $\sum\limits_{i=1}^{N} (X_i + a)$, when $a = 3$.

The sum of the values of a variable with a constant subtracted from each value is equal to the sum of the values of the variable minus N times the constant.

$$\sum_{i=1}^{N} (X_i - a) = \sum_{i=1}^{N} X_i - Na$$

Examples

Given the following values:

$$X_1 = 4, \qquad X_2 = 6, \qquad X_3 = 5.$$

Subtract a constant $a = 2$ from each value.

$$\sum_{i=1}^{N} (X_i - a) = 15 - 3(2)$$
$$= 9$$

Subtract a constant $a = 5$ from each value.

$$\sum_{i=1}^{N} (X_i - a) = 15 - 3(15)$$
$$= 0$$

Do Exercises 16 through 18 at the right.

QUALITATIVE AND QUANTITATIVE VARIABLES

There are two broad types of variables with which researchers work—qualitative and quantitative variables. Qualitative variables are ones which differ in kind, rather than by "how much." They are classified according to the attributes by which they differ, rather than in terms of the degree to which they share a given attribute. In contrast, a quantitative variable is one which expresses a difference in terms of *how much* individuals or objects possessing an attribute differ from one another.

Qualitative Variables

Qualitative variables are the result of classifying individuals or objects that share an attribute in common as "the same" and those that do not share this attribute as "different."

Examples

An American is anyone holding United States citizenship. A person holding citizenship in another country is not a United States citizen. (This is true for the overwhelming majority of people; however, there are persons who hold dual citizenship.)

Given the following values of the variable X:

$$X_1 = 4, \quad X_2 = 7, \quad X_3 = 2, \quad X_4 = 0,$$
$$X_5 = 9, \quad X_6 = 10, \quad X_7 = 1, \quad X_8 = 3.$$

16. Find $\sum_{i=1}^{N} (X_i - a)$, when $a = 2$.

17. Find $\sum_{i=1}^{N} (X_i - a)$, when $a = 7$.

18. Find $\sum_{i=1}^{N} (X_i - a)$, when $a = 4$.

A red automobile is different from a green automobile.

With respect to the attribute in question, qualitative variables are either–or.

Examples

A person is either a male or a female.

An animal is either a dog, or a cat, or an elephant, or. . . .

An automobile is either a Ford, or a Chevrolet, or a Fiat, or a Toyota, or. . . .

Do Exercises 19 through 24 at the right.

Data obtained with qualitative variables involves counting or enumeration.

Examples

The statistics class contains 15 male students and 22 female students.

The number of voters in Yasu City registered as Republican is 8,245; as Democrat, 7,993; as Independent, 2,372.

Do Exercises 25 through 27 at the right.

Nominal and Ordinal Scales

Nominal scales are qualitative variables in which no order, direction, or magnitude is implied. Ordinal scales are qualitative variables in which some order, direction, or magnitude is implied.

Examples

Car color is a nominal scale; rank in military (Private through General) is ordinal.

Political affiliation is a nominal scale; order of finish in "Most valuable player award" is ordinal.

Do Exercises 28 through 30 at the right.

Classification into ordinal scales involves algebra of inequalities.

Examples

The expression $a > b$ may mean: a is greater than b, higher rank than b, more prestigious than b, prettier than b, etc.

Check the qualitative variable in each pair listed below.

19. Sex; temperature

20. Height; color

21. Zip code; weight

22. Distance; hair color

23. Speed of travel; religious preference

24. Time; occupation

Check the member of each pair (italicized) which involves qualitative data.

25. Number of *blonds* in class; *height* of the tallest person in class

26. Number of *Datsuns* sold in Yasu City during 1976; the *weight* of a Datsun B-210

27. The *low temperature* in Gunnison, Colorado on January 5, 1977; the population of each of the *original thirteen colonies* at year's end in 1776

Select the member of each pair which involves ordinal scales.

28. Rank in leadership qualities; hair color

29. Worse natural disaster in history; make of automobile

30. Post position in horse races; type of animal

The expression $a < b$ may mean a is less than b, lower rank than b, less prestigious than b, etc.

Do Exercises 31 through 36 at the right.

The most common descriptive statistics with nominal and ordinal variables involve proportions and percentages.

Examples

The proportion of registered voters in Yasu City is 0.44.

The percentage of female students in the statistics class is 59.46.

Do Exercises 37 through 39 at the right.

Quantitative Variables

Quantitative variables indicate how much of a given attribute an object or person has.

Examples

Frances G. is 63 inches tall.

John McMahon weighs 235 pounds.

Quantitative variables can be added, subtracted, multiplied and divided.

Examples

Frances G. is 63 inches tall; Morris F. stands 70 inches high. Their combined height is 133 inches.

Sirloin roasts in the meat counter at the local supermarket weigh 56, 72, 35, and 46 ounces. The first two weigh 47 ounces more than the last two.

Do Exercises 40 through 43 at the right.

Continuous vs. Discrete Measurements

Some quantitative variables increase or decrease in continuous gradations. No matter how accurately measurements are made, it is always possible to conceive of a more accurate measure. Weight can be 125, 125.01, 125.0063, 125.006347, etc. Such variables are referred to as *continuous*.

Using symbols $>$ or $<$, complete the following statements.

31. General _____ Private

32. Corporal _____ Sergeant

33. Leader _____ Follower

34. Third post position _____ eighth post position

35. Winner of beauty pageant _____ also ran

36. Player sent down to minor leagues _____ most valuable player on team

Select the member of each pair which is *most likely* to involve proportions or percentages.

37. Mathematics scores of subjects following administration of a drug; cars sold in Yasu City by manufacturers

38. Number of horses during 1976 who finished first at each post position on an oval track; the heights of American males over 18 years of age

39. Salaries of employees in mining industries; number of different types of crimes during 1977

Indicate which member of each pair is a quantitative variable.

40. Number of blonds in class; height of tallest person in class

41. Number of Datsuns sold in Yasu City during 1976; the weight of a Datsun B-210

42. The low temperature in Gunnison, Colorado on January 5, 1977; the population of the original thirteen colonies at year's end in 1776

43. The size of the family; the colors of the rainbow

Examples

Height, weight, speed, temperature, altitude.

Other quantitative variables increase by *discrete*, rather than by continuous, amounts. A family unit may contain 0, 1, 2, 3, etc., children but not 1.43 children.

Do Exercises 44 through 47 at the right.

Descriptive statistics with quantitative variables commonly involve arithmetic procedures such as adding, subtracting, multiplying, and dividing.

Examples

The individual weights of four sixth-grade students are 80, 92, 73, and 75 pounds. Their mean weight ($\bar{X}$) is

$$\bar{X} = \frac{\sum\limits_{i=1}^{N} X_i}{N} = \frac{320}{4} = 80.$$

The range of weights (highest weight minus lowest weight) is

$$92 - 73 = 19.$$

Do Exercises 48 through 50 at the right.

PROPORTIONS AND PERCENTAGES

We previously noted that the descriptive statistics most commonly used with qualitative data involve proportions and percentages.

Distribution Ratio

A distribution ratio is defined as the ratio of a part to a total which includes that part. When a proportion is used as a distribution ratio, it can vary only between 0.00 and 1.00. If a equals the number of males in a class and b equals the number of females, the proportion of males is

$$\frac{a}{a + b}.$$

The proportion of females is

$$\frac{b}{a + b}.$$

Circle C (continuous) or D (discrete) for each variable shown below.

44. A child's spinner   C   D

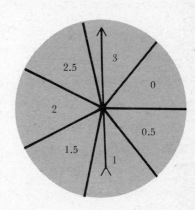

45. Weight of each horse at the County Fair   C   D

46. Size of family unit   C   D

47. The heights of various skyscrapers throughout the world   C   D

Select from each pair the one that is most likely to involve arithmetic procedures in calculating descriptive statistics.

48. Heights of various athletes; types of cars manufactured in Detroit

49. The number of children per family unit; eye color of students in statistics class

50. Number of winners at each post position on an oval track; daily miles driven to work by commuters

Examples

If there are no males in a class of 50 students, the proportion of males is

$$\frac{0}{50} = 0.00.$$

The proportion of females is

$$\frac{50}{50} = 1.00.$$

In 1974, the following numbers of firearms, in thousands, were produced domestically:

handguns = 1,715   rifles = 2,099

shotguns = 1,825

The proportion of handguns is

$$\frac{1715}{1715 + 2099 + 1825} = 0.30.$$

The proportion of rifles is

$$\frac{2099}{1715 + 2099 + 1825} = 0.37.$$

The proportion of shotguns is

$$\frac{1825}{1715 + 2099 + 1825} = 0.32.$$

Do Exercises 51 through 53 at the right.

The sum of all proportions in a distribution ratio is 1.00, although slight discrepancies may result from rounding. The proportions of handguns, rifles, and shotguns above are

$$0.30 + 0.37 + 0.32 = 0.99.$$

A proportion can be transformed into a percentage by multiplying the proportion by 100. The sum of all percentages in a distribution ratio is 100, although slight discrepancies may result from rounding.

Examples

The percentages of various types of firearms produced in 1973 are:

handguns, 30%   rifles, 37%

shotguns, 32%

Do Exercises 54 through 56 at the right.

Find the distribution ratios, expressed as a proportion, for each of the following.

51. In 1973, there were 15,840 male suicides and 4,625 female suicides. Find the proportion of male and female suicides (round to the second decimal).

52. In 1973, the deaths of 17,123 murder victims were caused by: guns, 11,249; cutting or stabbing, 2,985; strangulations or beatings, 1,445; blunt object, 848; arson, 173; other, 423. Find the proportions attributable to each cause of death (round to the second decimal).

53. In 1973, there were 20,465 suicides and 17,123 homicides. Find the proportion of the total attributable to suicide; the proportion attributable to homicide (round to the fourth decimal place).

Find the distribution ratio, expressed as a percentage, for each of the following.

54. The male and female suicides shown in Exercise 51 (round to the second decimal place).

2

Interclass Ratios

An interclass ratio is defined as a ratio of a part in a total to another part in the same total.

Examples

The ratio of male to female murders in 1973 was

15,840:4,625 = 3.42:1.

The ratio of rifles to handguns manufactured domestically in 1973 is

2099:1715 = 1.22:1.

Do Exercises 57 through 58 at the right.

Time Ratio or Time Relative

A time ratio is a measure which expresses the change in a series of values arranged in a time sequence and is typically shown as a percentage.

There are two main classes of time ratios:
1) those employing a fixed-base period and
2) those employing a moving base.

Fixed-base Time Ratio

The table below shows the number of male victims of homicide between the years 1968 and 1973.

Year	Number of male victims of homicide	Time ratio (fixed-base 1969)
1968	5,106	97.91
1969	5,215	100.00
1970	5,865	112.46
1971	6,455	123.78
1972	6,820	130.78
1973	7,411	142.11

The following steps are involved in constructing a fixed-base time ratio:

A. Select a given year as the fixed-base year (we chose 1969 for the present example).
B. Divide the number of homicides for each year, including 1969, by the number of homicides for the base year.
C. Multiply each answer by 100 to express as a percentage. Note that the fixed-base year is 100%.

---

55. The cause of death among murder victims shown in Exercise 52 (round to the second decimal place).

56. The percentage of violent deaths due to causes indicated in Exercise 53 (round to the nearest percentage).

Express the following as interclass ratios.

57. In 1973, the number of male deaths per 100,000 population attributed to heart disease was 415.2. The number of deaths due to cancer was 187.1 (round to the second decimal).

58. In 1973, the number of female deaths per 100,000 population attributable to heart disease was 309.1. The number of deaths due to cancer was 148.4 (round to the second decimal).

A value less than 100 means a lower rate than the base year; a value above 100 means a higher rate.

Fixed-based time ratios are widely used as economic indicators.

Examples

Cost of Living Index, Wholesale Price Index, Consumer Price Index, and Gross National Product

Do Exercises 59 and 60 at the right.

Moving-base Time Ratios

The most common type of moving-base time ratio uses the preceding year as the base.

Year	Number of male victims of homicide	Moving-base time ratio (preceding year)
1968	5,106	———
1969	5,215	102.13
1970	5,865	112.46
1971	6,455	110.06
1972	6,820	105.65
1973	7,411	108.67

Use the number of male homicides between the years 1968 and 1973 as the basic data and the following steps to construct a moving-base time ratio.

A. Divide the number of homicides for a given year by the number of homicides for the preceding year.

B. Multiply each answer by 100 to express as a percentage. Note that 1968 is left blank since there are no data for the preceding year (1967).

Moving-base time ratios are widely used to assess *rates* of change from year to year and as economic indicators to whether or not the *rate* of growth or decay is changing.

Examples

Cost of Living Index, Wholesale Price Index, Consumer Price Index, and Gross National Product

Do Exercises 61 and 62 at the right.

Prepare fixed-base time ratios for the following sets of data.

59.

Year	Number of female victims of homicide	Time ratio (fixed-base 1969)
1968	1,700	
1969	1,801	
1970	1,938	
1971	2,106	
1972	2,156	
1973	2,575	

60.

Year	U.S. population (in thousands)	Time ratio (fixed-base 1969)
1968	199,312	
1969	201,306	
1970	203,806	
1971	206,212	
1972	208,230	
1973	209,844	

Prepare moving-base time ratios for the following sets of data.

61.

Year	Number of female victims of homicide	Moving-base time ratio (preceding year)
1968	1,700	
1969	1,801	
1970	1,938	
1971	2,106	
1972	2,156	
1973	2,575	

62.

Year	U.S. population (in thousands)	Moving-base time ratio (preceding year)
1968	199,312	
1969	201,306	
1970	203,806	
1971	206,212	
1972	208,230	
1973	209,844	

CHAPTER 2 TEST

1. Given 7 quantities, X_1 through X_7, the correct notation for adding quantities 2 through 7 is:

   a) $\sum_{i=7}^{2} X_i$ b) $\sum_{i=7}^{1} X_i$

   c) $\sum_{i=2}^{N} X_i$ d) $\sum_{i=N}^{1} X_i$

2. Given: $X_1 = 2$, $X_2 = 9$, $X_3 = 0$, $X_4 = 1$, $\sum_{i=1}^{3} X_i$

   a) 4   b) 12   c) 11   d) 13

3. The symbolic notation $\sum_{i=2}^{N} X_i$ tells us to:

   a) add all quantities from X_1 through X_N

   b) add all quantities from $X = 2$ through X_N

   c) add all quantities from $X = 2$ through $X = N$

   d) add all quantities from X_2 through X_N

4. $\sum_{i=1}^{N} (X_i - a)$ equals:

   a) $\sum_{i=1}^{N} X_i - Na$ b) $NX_1 - Na$ c) $\sum_{i=1}^{N} X_i(-a)$ d) $\sum_{i=1}^{N} X_i - a$

Use the following values of the variable to answer Problems 5 through 8.

$X_1 = 3$, $X_2 = 1$, $X_3 = 14$, $X_4 = 2$, $X_5 = -10$, $X_N = 4$

5. Find $\sum_{i=3}^{4} (X_i + a)$, when $a = 1$.

   a) 16   b) 11   c) 13   d) 18

6. Find $\sum_{i=2}^{5} (X_i + a)$, when $a = 2$.

   a) 15   b) 13   c) 9   d) 25

7. Find $\sum_{i=1}^{N} (X_i + a)$, when $a = 3$.

   a) 17   b) 26   c) 41   d) 32

8. Find $\sum_{i=3}^{N} (X_i + a)$, when $a = 5$.

   a) 44   b) 30   c) 40   d) 19

Use the following values of the variable to answer Problems 9 through 12.

$X_1 = 2, X_2 = 8, X_3 = -6, X_4 = 1, X_N = 0$

9. Find $\sum\limits_{i=1}^{N} (X_i - a)$, when $a = 2$.

    a) 0   b) 15   c) 3   d) −5

10. Find $\sum\limits_{i=1}^{4} (X_i - a)$, when $a = 1$.

    a) −1   b) 0   c) 4   d) 9

11. Find $\sum\limits_{i=4}^{N} (X_i - a)$, when $a = -3$.

    a) −14   b) −5   c) −10   d) 16

12. Find $\sum\limits_{i=2}^{4} (X_i - a)$, when $a = 0$.

    a) 0   b) 3   c) 5   d) −3

13. Which of the following is a qualitative variable?

    a) species of snake   b) height of elm trees

    c) amount of water in a bath tub

    d) amount of snowfall in Gunnison, Colorado on a particular date

14. Which of the following is a quantitative variable?

    a) color of race horses   b) age of antiques

    c) rated attractiveness of a painting

    d) none of the preceding

15. Which of the following represents a nominal scale?

    a) temperature of water   b) relative humidity

    c) type of vegetable

    d) density of the atmosphere at various levels.

16. Which of the following represents an ordinal scale?

    a) make of sewing machine   b) telephone number

    c) price of shares of stock on a stock exchange

    d) ratings of emotional stability of military recruits

17. Which descriptive statistics are most likely to be used with nominally and ordinally scaled data?

    a) means   b) proportions

    c) percentages   d) both b and c

18. In which of the following are we most likely to employ proportions and percentages as descriptive statistics?

    a) daily temperatures   b) suicide data

    c) height of newborn infants

    d) barometric pressure   b) size of family unit

19. Which of the following represents continuous measurement?

    a) hair color

    c) number of hits in a baseball game

    d) weight

20. As opposed to simple counting, with which of the following are we most likely to employ arithmetic procedures such as adding, subtracting, multiplying and dividing?

    a) crime data

    b) employment and unemployment data

    c) daily temperature data in a given city

    d) rank order of finish in an Academy Award election.

Use the following data to complete Problems 21 through 24. Given: number of appliances sold by individuals during a given month:

Fran, 92; George, 88; Oscar, 67; Florence, 53

21. The distribution ratio of Fran's sales, expressed as a percentage is:

    a) 30.67   b) 0.3067   c) 3.26:1   d) 0.3261

22. The distribution ratio of Oscar's sales, expressed as a proportion is:

    a) 0.2876   b) 28.76   c) 0.2233   d) 22.33

23. The interclass ratio, George to Florence, is:

    a) 1.66:1   b) 0.60:1   c) 0.29:1   d) 1.26:1

24. The interclass ratio, Fran to Oscar, is:

    a) 0.73:1   b) 30.67:1   c) 0.22:1   d) 1.37:1

The following table shows the number of female suicides in the United States between 1968 and 1973. Use this table to complete Problems 25 through 30.

Year	Number of female suicides	
	White	Nonwhite
1968	5692	301
1969	6152	355
1970	6468	383
1971	6775	457
1972	6788	448
1973	6589	421

25. Using 1971 as the fixed-base year, the time ratio for nonwhites in 1969 is:

    a) 77.68   b) 128.73   c) 151.83   d) 65.86

26. Using 1973 as the fixed-base year, the time ratio for whites in 1968 is:

    a) 115.76   b) 97.07   c) 86.39   d) 83.85

27. Using 1968 as the fixed-base year, the time ratio for nonwhites in 1972 is:

    a) 1.49   b) 148.84   c) 67.19   d) 0.6719

28. The moving-base time ratio for nonwhites in 1970 is:

    a) 127.24   b) 107.89   c) 83.81   d) 119.32

29. The moving-base time ratio for whites in 1970 is:

    a) 105.14   b) 95.11   c) 113.63   d) 95.47

30. The moving-base time ratio for nonwhites in 1972 is:

    a) 106.41   b) 102.01   c) 148.84   d) 98.03

Practical

The following table shows the number of male suicides in the United States between 1968 and 1973. Use this table to complete Problems 31 through 35.

Year	Number of male suicides	
	Whites	Nonwhites
1968	14,520	859
1969	14,886	971
1970	15,591	1038
1971	15,802	1058
1972	16,476	1292
1973	16,823	1285

31. Using 1969 as the base year, calculate fixed-base time ratios for both white and nonwhite male suicides.

32. Express the suicide data for both white and nonwhite males as a moving-base time ratio.

33. Determine the distribution ratios (expressed as percentages) for white male suicides during the years 1968 through 1973 (round to the second decimal place).

34. Determine the distribution ratios (expressed as proportions) for nonwhite male suicides during the years 1968 through 1973 (round to the fourth decimal place).

35. Express the interclass ratio for each year, 1968 through 1973, of white male to nonwhite male suicides.

Frequency
Distributions

ORGANIZING DATA FROM QUANTITATIVE VARIABLES

When you collect data (often referred to as scores) on quantitative variables, you usually wind up with a chaotic and unorganized jumble of numerical values. The first thing you must do is organize the data in some coherent fashion. The frequency distribution is one way of accomplishing this goal.

Below are shown the Environmental Protection Agency's estimates of the city miles per gallon obtained by sixty 1976 model cars.

16	11	21	11	19	19
16	16	16	11	19	16
11	19	13	15	15	23
24	15	11	15	13	24
18	11	10	12	11	18
23	10	21	10	16	15
17	13	17	27	10	17
12	11	21	16	10	21
12	29	15	13	22	12
13	21	12	11	20	20

Constructing a Frequency Distribution

Find the highest mpg score among the sixty data points—this is 29 miles per gallon.

Prepare a column of numerical values headed by 29 and ending at the lowest miles-per-gallon score (10). Since the mpg values are given to the nearest mile, the entries should include all whole numbers (integers) between 29 and 10.

mpg	f	mpg	f
29		19	
28		18	
27		17	
26		16	
25		15	
24		14	
23		13	
22		12	
21		11	
20		10	

Do Exercise 1 at the right.

Now refer to the original scores. Find the first entry in the upper left hand corner (16) and place a slash alongside the score of 16. Proceed in like manner from score to score

OBJECTIVES

- Know how to construct an ungrouped frequency distribution from a collection of numerical data.
- Know how to construct a grouped frequency distribution.
- Know how to determine the true limits of a class interval.
- Know how to construct and interpret a cumulative frequency distribution.
- Know how to construct and interpret a cumulative percentage distribution.

EXERCISES

Shown below are the city miles-per-gallon ratings of thirty 1976 model cars.

16	13	11
11	10	17
24	21	12
16	17	10
12	13	16
11	28	12
11	21	19
19	19	15
15	15	13
11	12	12

1. Prepare a column headed by the highest value and ending at the lowest score. Include all integers between the two extreme scores.

until all scores have been recorded in this way. You have constructed a frequency distribution. The *f* at the head of the columns indicates the frequency of each score, i.e., the total number of times each score occurred.

mpg	*f*	mpg	*f*
29	1	19	1111
28		18	11
27	1	17	111
26		16	1111111
25		15	111111
24	11	14	
23	11	13	11111
22	1	12	11111
21	11111	11	111111111
20	11	10	11111
		$\Sigma f = 60$	

Note that the sum of the frequencies, Σf, equals N.

Do Exercise 2 at the right.

Constructing Grouped Frequency Distributions

Inspection of the preceding frequency distribution reveals that the scores are rather widely spread and a number of scores have zero frequency associated with them. It is customary under such circumstances to group the scores into class intervals. There is no generally agreed upon convention for deciding the most desirable number of class intervals for a given set of data. However, there is broad agreement tha⁺ most data can be handled well with between 10 and 20 class intervals.

For the case of the sixty preceding scores, we shall strive for about ten class intervals. The procedures are as follow:

A. Find the difference between the highest and lowest real values contained in the original data. Add 1 to obtain the total number of scores or potential scores. In the present example, the result is $(29 - 10) + 1 = 20$.

B. Divide this figure by 10 to obtain the number of scores or potential scores in each interval. Round to the nearest whole number since the scores are expressed as integers. In the present example, the width of the interval, i, is

$$\frac{20}{10} = 2.$$

2. Prepare a frequency distribution of the scores shown in Exercise 1.

mpg	*f*	mpg	*f*
28		18	
27		17	
26		16	
25		15	
24		14	
23		13	
22		12	
21		11	
20		10	
19			

C. Take the lowest score in the original data as the minimum value in the lowest class interval. Add to this $i - 1$ to obtain the maximum score in the lowest interval. In the present example, the lowest class interval is 10–11.

D. The next higher class interval begins at the integer following the maximum score of the lower class interval. In the present example, the next integer is 12. Follow these procedures for each successive higher class interval until all the scores are included in their appropriate class intervals.

E. Assign each obtained score to the class interval within which it is included.

The table below shows the grouped frequency distribution based upon the miles-per-gallon data of sixty 1976 automobiles.

Class interval	f
28–29	1
26–27	1
24–25	2
22–23	3
20–21	7
18–19	6
16–17	10
14–15	6
12–13	10
10–11	14
	$N = 60$

Do Exercises 3 through 5 at the right.

THE TRUE LIMITS OF A CLASS INTERVAL

Suppose you weigh several cuts of meat on a scale that is accurate to the nearest ounce, and you record the following readings:

43 oz.   40 oz.   6 oz.   25 oz.

The true limits of each weight are equal to that weight plus and minus one half the unit of measure. Since the unit of measure is in ounces, the true limits of each of the above weights are

43 ± 0.5 oz.   or   42.5–43.5 oz.

40 ± 0.5 oz.   or   39.5–40.5 oz.

6 ± 0.5 oz.   or   5.5– 6.5 oz.

3. Construct a grouped frequency distribution of the thirty mpg scores, using $i = 2$.

4. Construct a grouped frequency distribution of the thirty mpg scores, using $i = 3$.

5. Construct a grouped frequency distribution of the thirty mpg scores, using $i = 5$.

If a measure is accurate to the tenths place, the true limits of a number are that number ± 0.05.

Examples

True limits of 6.3 are 6.25–6.35.

True limits of 654.7 are 654.65–654.75.

Generalization: The true limits of any number are that number plus and minus one half the unit of measure.

Example

If measures are accurate to the nearest hundred, 65,600 is between 65,550 and 65,650.

Do Exercises 6 through 8 at the right.

The lower real limit of an interval is the lower apparent limit minus one half the unit of measure.

Example

In the grouped frequency distribution of the mpg scores of sixty 1976 automobiles, the apparent and real lower limits of each interval are as follow:

Apparent lower limit	Real lower limit
28	27.5
26	25.5
24	23.5
22	21.5
20	19.5
18	17.5
16	15.5
14	13.5
12	11.5
10	9.5

Do Exercises 9 through 16 at the right.

The upper real limit of an interval is the upper apparent limit plus one half the unit of measure.

Example

In the grouped frequency distribution of the sixty mpg scores, the apparent and real upper limits of each interval are as follow.

Write the true limits of each of the following numbers.

6. The unit of measure is in the units place.
   a) 16
   b) 12
   c)   0
   d) 100

7. The unit of measure is in the tens place.
   a) 160
   b) 190
   c) 195
   d) 870

8. The unit of measure is in the tenths place.
   a) 65.5
   b) 193.0
   c) 100.1
   d) 99.9

Write the lower real limit of each of the following class intervals. The unit of measure is in the units place.

9. 10–12

10. 19–21

11. 26–28

12. 5–9

13. 0–4

14. 16–23

15. 100–110

16. 60–69

Apparent upper limit	Real upper limit
29	29.5
27	27.5
25	25.5
23	23.5
21	21.5
19	19.5
17	17.5
15	15.5
13	13.5
11	11.5

Do Exercises 17 through 24 at the right.

CUMULATIVE FREQUENCY DISTRIBUTION

A cumulative frequency distribution is obtained by successively summing or cumulating the frequency counts from the lowest class interval to the highest. (It is also possible to cumulate in the opposite direction but that is a special case beyond the scope of this book.)

Example

Shown below are the grouped frequency and cumulative frequency distributions of the sixty miles-per-gallon scores.

Class interval			
Apparent limits	Real limits	f	Cum f
28–29	27.5–29.5	1	60
26–27	25.5–27.5	1	59
24–25	23.5–25.5	2	58
22–23	21.5–23.5	3	56
20–21	19.5–21.5	7	53
18–19	17.5–19.5	6	46
16–17	15.5–17.5	10	40
14–15	13.5–15.5	6	30
12–13	11.5–13.5	10	24
10–11	9.5–11.5	14	14
		$\Sigma f = 60$	

Do Exercises 25 through 27 at the right.

Each entry in the cumulative frequency distribution indicates the number of all cases or frequencies below the upper real limit of its corresponding interval.

Write the upper real limit of each of the following class intervals. The unit of measure is in the units place.

17. 10–12

18. 19–21

19. 26–28

20. 5–9

21. 0–4

22. 16–23

23. 100–110

24. 60–69

Construct cumulative frequency distributions based upon:

25. Exercise 3

26. Exercise 4

27. Exercise 5

Examples

The 59 corresponding to the interval 25.5–27.5 indicates that fifty-nine of the sixty frequencies are below 27.5.

The 14 corresponding to the interval 9.5–11.5 indicates that fourteen of the sixty frequencies fall below 11.5.

Do Exercises 28 through 32 at the right.

CUMULATIVE PERCENTAGE DISTRIBUTION

To transform a cumulative frequency distribution to a cumulative percentage distribution, divide each entry in the cumulative f column by N and multiply by 100.

Example

| Class interval | | | | |
Apparent limits	Real limits	f	Cum f	Cum %
28–29	27.5–29.5	1	60	100
26–27	25.5–27.5	1	59	98
24–25	23.5–25.5	2	58	97
22–23	21.5–23.5	3	56	93
20–21	19.5–21.5	7	53	88
18–19	17.5–19.5	6	46	77
16–17	15.5–17.5	10	40	67
14–15	13.5–15.5	6	30	50
12–13	11.5–13.5	10	24	40
10–11	9.5–11.5	14	14	23
	$\Sigma f = 60$			

Do Exercises 33 through 35 at the right.

Each entry in the cumulative percentage distribution indicates the number of all cases or frequencies below the upper real limit of its corresponding interval.

Examples

The 100% corresponding to the interval 27.5–29.5 indicates that 100% of the frequencies are below 29.5. The percent above 29.5 is zero. The 50% corresponding to the interval 13.5–15.5 indicates that 50% of the frequencies fall below 15.5. The percent of cases above 15.5 is 50.

Do Exercises 36 through 41 at the right.

Answer the following based on the cumulative frequency graph of sixty mpg scores.

28. The score below which fifty-eight cases fall is _____.

29. The score below which thirty cases fall is _____.

30. The score below which twenty-four cases fall is _____.

31. The score below which fifty-three cases fall is _____.

32. The score below which forty cases fall is _____.

Construct cumulative percentage distributions from:

33. Exercise 25

34. Exercise 26

35. Exercise 27

Using the cumulative percentage distribution of the sixty mpg scores, indicate the percent of cases falling below and above the following values.

36. 21.5

37. 25.5

38. 13.5

39. 11.5

40. 9.5

41. 17.5

CHAPTER 3 TEST

Use the following set of scores to complete Problems 1 through 4.

4.8	3.0	4.5	1.2	2.8
1.5	3.2	3.0	4.3	4.4
2.8	1.8	2.5	3.5	4.6
1.6	2.9	2.0	2.7	4.0
2.9	4.2	2.8	1.4	1.8
2.3	2.6	3.1	3.4	3.7
2.7	2.8	2.9	3.6	2.6
3.5	3.2	3.4	2.0	2.1
2.9	3.9	2.8	1.7	2.5
2.6	3.4	3.7	3.0	2.7
2.5	3.3	2.2	2.3	2.8

1. In constructing an ungrouped frequency distribution, the lowest score is:

   a) 1.3   b) 1.2   c) 1.1   d) 1.4

2. In constructing an ungrouped frequency distribution, the highest score is:

   a) 4.8   b) 4.5   c) 5.5   d) 4.6

3. The N is:

   a) 50   b) 60   c) 55   d) 5

4. In the above set of scores, the true limits of a score of 2.2 are:

   a) 2.1–2.3   b) 2.19–2.21   c) 2.15–2.25   d) 2.15–2.21

Use the following frequency distribution of scores to complete Problems 5 through 10.

X	f	X	f	X	f	X	f
5.3	1	4.3	0	3.4	4	2.5	2
5.2	0	4.2	2	3.3	6	2.4	0
5.1	1	4.1	1	3.2	3	2.3	2
5.0	1	4.0	2	3.1	3	2.2	1
4.9	1	3.9	3	3.0	2	2.1	1
4.8	1	3.8	1	2.9	1	2.0	1
4.7	1	3.7	2	2.8	2	1.9	1
4.6	0	3.6	1	2.7	1	1.8	0
4.5	1	3.5	3	2.6	1	1.7	1
4.4	1						

5. If we prepared a grouped frequency distribution, $i = 0.3$, the apparent limits of the lowest class interval would be:

   a) 1.5–1.7   b) 1.55–1.65   c) 1.5–1.8   d) 1.7–1.9

6. The true limits of the interval 3.2–3.4 are:

   a) 3.1–3.5   b) 3.25–3.45   c) 3.1–3.5   d) 3.15–3.45

3

7. The frequency included in the interval 3.8–4.0 is:

   a) 0.6   b) 6   c) 5   d) 0.3

8. The frequency included in the interval 4.7–4.9 is:

   a) 0.3   b) 3   c) 5   d) 2

9. If we prepared a grouped frequency distribution, $i = 0.5$, how many class intervals would we have?

   a) 5   b) 8   c) 11   d) 10

10. If $i = 0.5$, the upper real limit of the lower class interval is:

   a) 31.5   b) 32.5   c) 33.5   d) 2.15

11. If we are measuring to the nearest 10 pounds, the true limits of a weight reported as 48,560 pounds are:

   a) 48,555–48,565

   b) 48,560–48,570

   c) 48,555–48,561

   d) 48,559–48,561

12. If measuring to the nearest hundredths of a pound, the true limits of a weight reported as 15.45 are:

   a) 15.40–15.50

   b) 15.449–15.451

   c) 15.445–15.455

   d) 15.35–15.55

13. If the apparent limits of a grouped frequency distribution are 6.95–7.04, the real limits are:

   a) 6.90–7.05

   b) 6.955–7.045

   c) 6.945–7.045

   d) 6.95- 7.05

Problems 14 through 21 are based on the grouped frequency distribution shown below.

Class interval	f
5.8–6.2	1
5.3–5.7	2
4.8–5.2	4
4.3–4.7	7
3.8–4.2	12
3.3–3.7	15
2.8–3.2	13
2.3–2.7	8
1.8–2.2	5
1.3–1.7	2
.8–1.2	1

14. The width of the class interval, i, is:

    a) 0.5   b) 0.4   c) 5   d) 4

15. The apparent limits of the class interval containing the sixty-seventh score (counting from the bottom) are:

    a) 5.25–5.75   b) 3.8–4.2   c) 5.8–6.2   d) 4.8–5.2

16. The apparent limits of the class interval containing the twenty-fourth score are:

    a) 2.3–2.7   b) 2.8–3.2   c) 2.75–3.25   d) 2.25–2.75

17. The real limits of the interval containing the eighth score are:

    a) 1.25–1.75   b) 2.25–2.75   c) 1.75–2.25   d) 2.3–2.7

18. The real limits of the interval containing the fifteenth score are:

    a) 3.3–3.7   b) 3.25–3.75   c) 2.25–2.75   d) 2.75–3.25

19. The score below which sixty-three frequencies fall is:

    a) 4.75   b) 4.8   c) 4.7   d) 4.25

20. The score below which eight frequencies fall is:

    a) 2.75   b) 2.3   c) 2.25   d) 2.2

21. The score below which fifty-six frequencies fall is:

    a) 5.75   b) 4.25   c) 4.2   d) 4.3

Problems 22 through 26 are based on the following cumulative frequency distribution.

Apparent limits	Real limits	f	Cum f
6.3–6.8	6.25–6.85	1	73
5.7–6.2	5.65–6.25	3	72
5.1–5.6	5.05–5.65	5	69
4.5–5.0	4.45–5.05	9	64
3.9–4.4	3.85–4.45	12	55
3.3–3.8	3.25–3.85	17	43
2.7–3.2	2.65–3.25	13	26
2.1–2.6	2.05–2.65	7	13
1.5–2.0	1.45–2.05	4	6
.9–1.4	.85–1.45	2	2
		$N = 73$	

22. What percent of cases fall below a score of 2.65?

    a) 9.59   b) 3.63   c) 35.61   d) 17.81

23. What percent of cases fall below a score of 5.65?

    a) 94.52   b) 6.85   c) 7.74   d) 98.63

24. What percent of cases fall below a score of 3.25?

    a) 17.81   b) 35.62   c) 58.90   d) 23.29

25. What percent of cases fall below a score of 5.05?

    a) 6.92   b) 0.8767   c) 87.67   d) 94.52

26. What percent of cases fall below a score of 1.45?

    a) 1.99   b) 2.74   c) 8.22   d) 0.00

Practical

Use the following set of scores to complete Problems 27 through 30.

2.8	2.7	2.5	2.1	2.6	3.7
1.8	4.0	4.6	4.4	2.8	2.3
3.0	1.7	2.0	3.6	3.4	1.4
2.7	3.5	4.3	1.2	2.2	3.7
2.8	3.4	2.9	3.1	2.8	2.0
2.5	3.0	4.5	3.3	3.4	3.9
3.2	2.8	2.6	4.2	2.9	1.8
3.2	3.0	2.5	2.6	2.9	3.5
2.7	2.3	2.9	1.6	2.8	1.5

27. Construct an ungrouped frequency distribution of the scores.

28. Construct a grouped frequency distribution of the scores using $i = 3$.

29. Construct a cumulative frequency and cumulative percentage distribution of the grouped scores ($i = 3$).

30. a) The score corresponding to a cumulative percentage of 59.26 is _____.

    b) The score corresponding to a cumulative percentage of 87.04 is _____.

    c) The score corresponding to a cumulative percentage of 9.26 is _____.

    d) The score corresponding to a cumulative percentage of 16.67 is _____.

31. a) A score of 2.95 has a corresponding cumulative percentage of _____.

    b) A score of 1.45 has a corresponding cumulative percentage of _____.

    c) A score of 3.25 has a corresponding cumulative percentage of _____.

    d) A score of 4.15 has a corresponding cumulative percentage of _____.

Percentiles

PERCENTILE RANK OF A SCORE

The percentile rank of a score is the percent of frequencies in a comparison group which achieved scores lower than the one cited.

Examples

Referring back to the cumulative percentage distribution of sixty mpg scores, a score of 13.5 has a percentile rank of 40; a score of 19.5 has a percentile rank of 77.

In Chapter 3, when we determined the cumulative percent corresponding to the upper limit of each interval we were, in effect, calculating the percentile rank of the score at the upper real limit of the interval. However, it is often necessary to determine the percentile rank of scores within an interval. This can be accomplished in an approximate form by constructing a cumulative percentage graph.

Constructing a Cumulative Percentage Graph

A. Mark, on a piece of graph paper, equally separated divisions along the horizontal axis corresponding to the upper limit of each class interval.

B. Add one mark corresponding to the lower real limits of the lowest class interval. This locus will correspond to the zero point on the vertical, or cumulative percentage, axis.

C. Mark, on the vertical axis, a minimum of ten equally spaced divisions, each corresponding to ten percentage points on the cumulative percentage axis. (For greater accuracy, you can use a greater number of subdivisions. If you use 100 subdivisions, each one will correspond to a separate percentile rank.)

D. Plot the point at the upper real limit of each interval that corresponds to its cumulative percentage.

E. Join the points with a ruler and bring down to zero percentage at the lower real limit of the lowest class interval.

F. Label the axes as shown in the following cumulative percentage graph of the sixty mpg scores.

OBJECTIVES

*   Know how to approximate the percentile rank of a score by use of a cumulative percentage graph.
*   Know how to find the approximate score corresponding to a given percentile rank by use of a cumulative percentage graph.
*   Using interpolation techniques, be able to calculate the percentile ranks of scores.
*   Be able to calculate the score corresponding to a given percentile rank.

4

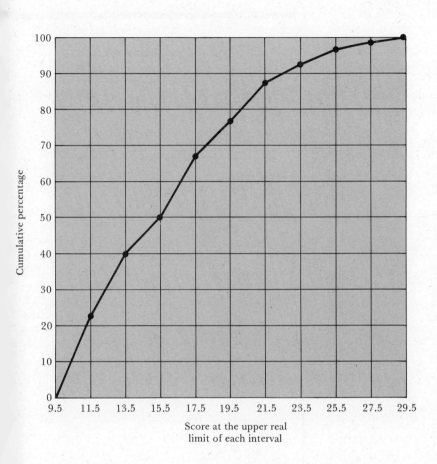

Cumulative percentage

Score at the upper real
limit of each interval

Do Exercises 1 through 3 at the right.

EXERCISES

Construct a cumulative percentage graph
based on:

1. Exercise 33, Chapter 3

2. Exercise 34, Chapter 3

3. Exercise 35, Chapter 3

Estimating Percentile Ranks Graphically

Let's say you want to estimate the percentile rank of a score of 16. Since 16 is a quarter of the distance (0.5/2 = 0.25) between 15.5 and 17.5, find a point on the horizontal axis approximately one quarter of the distance between the two real limits of that interval. Erect a vertical line. The point at which it intercepts the curve represents the approximate percentile rank of a score of 16. The figure below shows the means of estimating the percentile ranks of the following scores: 10, 12, 21, 28.

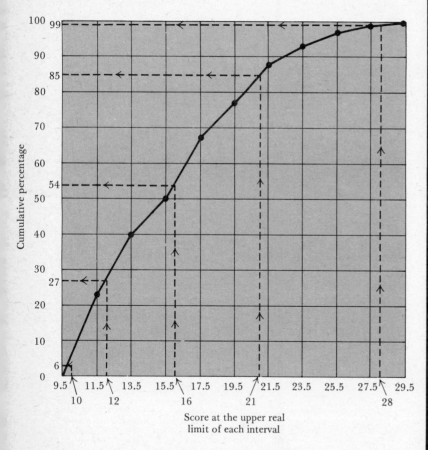

Score at the upper real limit of each interval

Do Exercises 4 through 17 at the right.

Estimate graphically the percentile ranks of the following scores.

4. 11

5. 14

6. 15

7. 17

8. 18

9. 19

10. 20

11. 22

12. 23

13. 24

14. 25

15. 26

16. 27

17. 29

4

Estimating Graphically the Score Corresponding to a Percentile Rank

There are occasions when a person is given the percentile rank of a score, and he or she wants to find the score corresponding to that percentile rank. To do so, locate along the vertical axis the percentile rank (cumulative percentage) of interest. Hold a straightedge perpendicular to this axis. The point at which it intercepts the cumulative percentage line represents the approximate score corresponding to the known percentile rank.

Examples

The score at the 30th percentile is approximately 12.

The score at the 75th percentile is approximately 19.

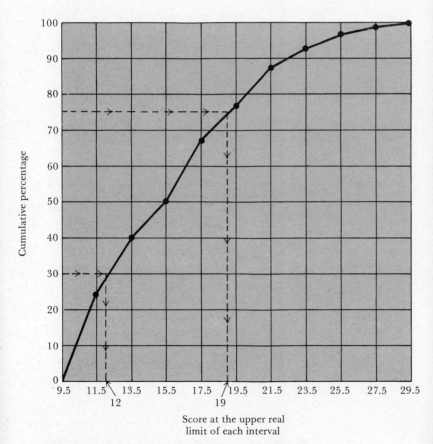

Score at the upper real
limit of each interval

Do Exercises 18 through 30 at the right.

Estimate the scores corresponding to the following percentile ranks.

18. 5

19. 10

20. 15

21. 25

22. 35

23. 40

24. 45

25. 55

26. 60

27. 70

28. 80

29. 90

30. 95

Calculating the Percentile Rank of a Score

For the precise determination of the percentile rank of a score, special procedures are required.

Example

Find the percentile rank of a mpg score of 13.

A. Construct a cumulative frequency distribution, using the steps outlined in Chapter 3 (repeated below).

B. Identify the interval containing that score. In the present example, this interval is 11.5–13.5.

C. Subtract the score at the lower real limit of the interval from the score in question: $13 - 11.5 = 1.5$.

D. Note the frequency within that interval: 10.

E. Divide the value found in step C by the width of the interval and multiply by the frequency found in step D. Thus,

$$\left(\frac{1.5}{2}\right) \times 10 = 7.5.$$

This represents the frequency within the interval 11.5–13.5 corresponding to a score of 13.

F. Note the cumulative frequency appearing in the *adjacent lower* interval: cum $f = 14$.

G. Add the value found in step E to the value found in step F. Thus, $7.5 + 14 = 21.5$.

H. Divide the value found in step G by N and multiply by 100. Thus,

$$\frac{21.5}{60} \times 100 = 35.83.$$

Interpretation: Approximately 64 percent of the cars in the sample obtained better than 13 mpg and about 36 percent obtained poorer mileage.

Real limits of class interval	f	cum f
27.5–29.5	1	60
25.5–27.5	1	59
23.5–25.5	2	58
21.5–23.5	3	56
19.5–21.5	7	53
17.5–19.5	6	46
15.5–17.5	10	40
13.5–15.5	6	30
11.5–13.5	10	24
9.5–11.5	14	14

4

Examples

Percentile rank of a score of 26 is

$$\frac{\left(\frac{0.5}{2}\right)1 + 58}{60} \times 100 = \frac{58.25}{60} \times 100$$

$$= 97.08.$$

Percentile rank of a score of 10 is

$$\frac{\left(\frac{0.5}{2}\right)14 + 0}{60} \times 100 = \frac{3.5}{60} \times 100$$

$$= 5.83.$$

Do Exercises 31 through 36 at the right.

Calculating the Score Corresponding to a Percentile Rank

Given a percentile rank, the following steps will permit you to determine the score corresponding to that rank. These procedures are valuable in checking out the accuracy of the prior calculations of the percentile rank of a score.

Given: Previously calculated percentile rank of 35.83. Find the score corresponding to this percentile rank. If our previous work was correct, we expect to obtain a score of 13.

A.   Multiply 35.83 by N and divide by 100. Thus,

$$\frac{35.83 \times 60}{100} = 21.50.$$

B.   Find the class interval containing the 21.50th cumulative frequency. This is the interval 11.5–13.5.

C.   Subtract from the value obtained in step A the cumulative frequency at the upper real limit of the *adjacent lower* interval, i.e., $21.50 - 14 = 7.5$. This value shows that the desired score is 7.5 frequencies within the interval 11.5–13.5.

D.   Divide the value found in step C by the frequency within the interval and multiply by the width of the interval ($i = 2$). Thus,

$$\frac{7.5}{10} \times 2 = 1.5.$$

E.   Add the value in step D to the score at the lower real limit of the interval containing the 21.50th frequency. Thus

$$11.5 + 1.5 = 13$$

is the answer.

Given below are the city mpg of the indicated cars. Determine the percentile rank of that car in mpg.

31. Chevette, 22 mpg

32. Hornet wagon, 16 mpg

33. Audi Fox, 24 mpg

34. Vega Kammback, 19 mpg

35. Cadillac Eldorado, 11 mpg

36. Toyota Corolla, 20 mpg

Examples

Given: Percentile rank is 97.08. Find the corresponding score.

A. $\dfrac{97.08 \times 60}{100} = 58.25$

B.   25.5

C. $58.25 - 58 = 0.25$

D. $\left(\dfrac{0.25}{1}\right) 2 = 0.50$

E. $25.5 + 0.5 = 26$

Given: Percentile rank is 5.83. Find the corresponding score.

A. $\dfrac{5.83 \times 60}{100} = 3.50$

B.   9.5

C. $3.50 - 0 = 3.50$

D. $\left(\dfrac{3.5}{14}\right) 2 = 0.50$

E. $9.5 + 0.5 = 10.0$

Do Exercises 37 through 42 at the right.

Find the scores corresponding to the following percentile ranks.

37. 10

38. 75

39. 25

40. 18

41. 84

42. 16

4

CHAPTER 4 TEST

Use the following cumulative percentage graph to complete Problems 1 through 8.

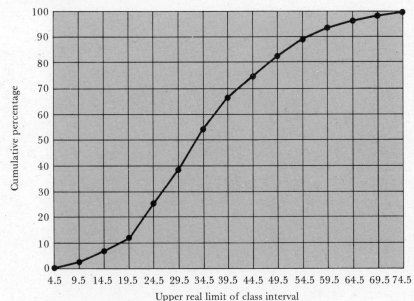

1. The approximate percentile rank of a score of 9 is:

   a) 5   b) 2   c) 16   d) 18

2. The approximate percentile rank of a score of 42 is:

   a) 70   b) 30   c) 33   d) 65

3. The approximate percentile rank of a score of 27 is:

   a) 22   b) 32   c) 25   d) 36

4. The approximate percentile rank of a score of 64 is:

   a) 40   b) 37   c) 93   d) 96

5. The approximate percentile rank of a score of 55 is:

   a) 90   b) 37   c) 86   d) 34

6. The score approximately corresponding to a percentile rank of 60 is:

   a) 32   b) 37   c) 94   d) 97

7. The score approximately corresponding to a percentile rank of 17 is:

   a) 9   b) 6   c) 22   d) 25

8. The score approximately corresponding to a percentile rank of 50 is:

   a) 30   b) 33   c) 82   d) 85

Problems 9 through 20 are based on the cumulative frequency distribution of scores shown below.

Apparent limits	Real limits	f	Cum f
150–154	149.5–154.5	2	110
145–149	144.5–149.5	2	108
140–144	139.5–144.5	3	106
135–139	134.5–139.5	5	103
130–134	129.5–134.5	7	98
125–129	124.5–129.5	9	91
120–124	119.5–124.5	9	82
115–119	114.5–119.5	13	73
110–114	109.5–114.5	17	60
105–109	104.5–109.5	14	43
100–104	99.5–104.5	12	29
95– 99	94.5– 99.5	4	17
90– 94	89.5– 94.5	5	13
85– 89	84.5– 89.5	5	8
80– 84	79.5– 84.5	3	3

$$N = 110$$

9. The percentile rank of a score of 127 is:

   a) 80.45   b) 26.59   c) 78.64   d) 86.82

10. The percentile rank of a score of 112 is:

   a) 62.27   b) 39.76   c) 55.22   d) 46.82

11. The percentile rank of a score of 134 is:

   a) 92.04   b) 94.82   c) 88.45   d) 85.65

12. The percentile rank of a score of 104 is:

   a) 25.27   b) 36.18   c) 28.07   d) 17.16

13. The percentile rank of a score of 83 is:

   a) 3.55   b) 0.82   c) 8.03   d) 5.30

14. The percentile rank of a score of 148 is:

   a) 99.45   b) 104.31   c) 97.45   d) 97.64

15. What score corresponds to a percentile rank of 96?

   a) 142.16   b) 146.06   c) 143.83   d) 148.83

16. What score corresponds to a percentile rank of 25?

   a) 103.88   b) 129.7   c) 108.88   d) 124.7

17. What score corresponds to a percentile rank of 50?

   a) 118.03   b) 155.3   c) 150.3   d) 113.03

18. What score corresponds to a percentile rank of 75?

   a) 125.4   b) 129.78   c) 124.78   d) 130.4

19. What score corresponds to a percentile rank of 84?

a) 131.46   b) 130.50   c) 136.46   d) 135.50

20. What score corresponds to a percentile rank of 16?

a) 104.75   b) 113.90   c) 118.90   d) 99.75

Practical

Problems 21 through 25 are based on the following cumulative percentage distribution of scores.

Class interval	Real limits of interval	f	Cum f	Cum %
41–43	40.5–43.5	1	98	100.00
38–40	37.5–40.5	2	97	98.98
35–37	34.5–37.5	4	95	96.94
32–34	31.5–34.5	9	91	92.86
29–31	28.5–31.5	10	82	83.67
26–28	25.5–28.5	13	72	73.47
23–25	22.5–25.5	18	59	60.20
20–22	19.5–22.5	14	41	41.84
17–19	16.5–19.5	11	27	27.55
14–16	13.6–16.5	8	16	16.33
11–13	10.5–13.5	5	8	8.16
8–10	7.5–10.5	2	3	2.06
5– 7	4.5– 7.5	1	1	1.02

21. Plot a cumulative percentage graph of the given data.

22. Estimate from the graph the cumulative percentage of the following scores.

a) 12   b) 24   c) 32   d) 36   e) 41

23. Find the scores approximating the following percentile ranks.

a) 20   b) 36   c) 54   d) 78   e) 94

24. Calculate the percentile ranks of the following scores.

a) 12   b) 24   c) 47   d) 73   e) 88

25. Calculate the scores corresponding to the following percentile ranks.

a) 20   b) 36   c) 32   d) 37   e) 41

Measures of Central Tendency

There are three measures of central tendency—the mean, the median, and the mode. They are referred to as measures of central tendency because they reflect numerical values in the central region of a distribution of scores. However, since each is defined in a different way, the three measures are often not in complete agreement.

UNGROUPED FREQUENCY DISTRIBUTIONS

The table below reproduces the ungrouped frequency distribution of the miles-per-gallon scores of sixty 1976 automobiles.

X	f	X	f
29	1	19	4
28	0	18	2
27	1	17	3
26	0	16	7
25	0	15	6
24	2	14	0
23	2	13	5
22	1	12	5
21	5	11	9
20	2	10	5
			$N = 60$

Calculating the Mean

The mean is defined as the sum of the scores divided by their number, that is,

$$\bar{X} = \frac{\Sigma X}{N} .$$

However, when we calculate the mean of scores in a frequency distribution—grouped or ungrouped—we take advantage of our knowledge that multiplication is successive addition. Thus, instead of adding the score of 11 nine times (see above) we multiply 11 by its corresponding frequency to obtain a sum of 99. In fact, we multiply each score by its corresponding frequency (see column fX below) prior to summing the fX column. For a frequency distribution, the mean is defined as

$$\bar{X} = \frac{\Sigma fX}{N} .$$

OBJECTIVES

- Be able to calculate the three measures of central tendency.
- Know how the mean, median, and mode are similar and how they differ.
- Know how to calculate a weighted mean.

5

1	2	3	1	2	3
X	f	fX	X	f	fX
29	1	29	19	4	76
28	0	0	18	2	36
27	1	27	17	3	51
26	0	0	16	7	112
25	0	0	15	6	90
24	2	48	14	0	0
23	2	46	13	5	65
22	1	22	12	5	60
21	5	105	11	9	99
20	2	40	10	5	50
					ΣfX = 956

$$\bar{X} = \frac{956}{60} = 15.93$$

A. Multiply each score (column 1) by its corresponding frequency (column 2) and record in column fX.

B. Sum column fX (column 3) to obtain ΣfX.

C. Divide by N to obtain mean.

Do Exercise 1 at the right.

Calculating the Median

The median is merely a special case of a percentile rank. It is, in fact, the score at the 50th percentile. In other words, it is the score that exactly divides a distribution into two halves—one half of the scores are above and one half are below the median. The following steps are used with an ungrouped frequency distribution.

A. Obtain a cumulative frequency distribution (see below).

B. Multiply 50 by N and divide by 100. Thus,

$$\frac{50 \times 60}{100} = 30.$$

C. Find the score that contains the 30th frequency within its real limits.

D. Subtract from the value found in step B the cumulative frequency at the upper real limit of the *adjacent lower* score, that is,

$$30 - 24 = 6.$$

EXERCISES

1. Find the mean of the following ungrouped frequency distribution of miles-per-gallon scores of thirty 1976 automobiles.

X	f	ΣfX	X	f	ΣfX
28	1		18	0	
27	0		17	2	
26	0		16	3	
25	0		15	2	
24	1		14	1	
23	0		13	4	
22	0		12	4	
21	2		11	5	
20	0		10	2	
19	3				

E. Divide the value found in step D by the frequency within the real limits of the score containing the 30th frequency, that is,

$$\frac{6}{6} = 1.00.$$

F. Add the value in step E to the lower real limit of the score containing the 30th frequency. Thus,

$$14.5 + 1 = 15.5$$

is the answer.

X	Real limits of X	f	Cum f
29	28.5–29.5	1	60
28	27.5–28.5	0	59
27	26.5–27.5	1	59
26	25.5–26.5	0	58
25	24.5–25.5	0	58
24	23.5–24.5	2	58
23	22.5–23.5	2	56
22	21.5–22.5	1	54
21	20.5–21.5	5	53
20	19.5–20.5	2	48
19	18.5–19.5	4	46
18	17.5–18.5	2	42
17	16.5–17.5	3	40
16	15.5–16.5	7	37
15	14.5–15.5	6	30
14	13.5–14.5	0	24
13	12.5–13.5	5	24
12	11.5–12.5	5	19
11	10.5–11.5	9	14
10	9.5–10.5	5	5

Do Exercise 2 at the right.

Determining the Mode

The mode is merely the score with the highest associated frequency. Since 11 miles per gallon occurs with the greatest frequency (9), the mode is a score of 11.

Do Exercise 3 at the right.

Look at the frequency polygon appearing below. Note that the distribution of scores is not symmetrical about the mean. There is piling up of scores at the lower end of the distribution and greater spread of scores at the higher end. The distribution is said to be *positively skewed* or skewed to the right.

2. Find the median of the ungrouped frequency distribution of miles-per-gallon scores of thirty 1976 automobiles.

X	f	Cum f	X	f	Cum f
28	1		18	0	
27	0		17	2	
26	0		16	3	
25	0		15	3	
24	1		14	0	
23	0		13	4	
22	0		12	4	
21	2		11	5	
20	0		10	2	
19	3				

3. Determine the mode of the ungrouped frequency distribution of the same thirty miles-per-gallon scores.

5

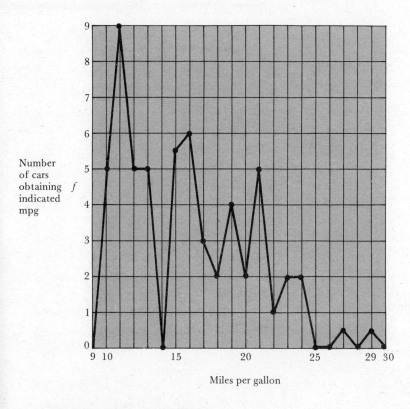

Miles per gallon

Do Exercise 4 at the right.

GROUPED FREQUENCY DISTRIBUTIONS

Calculating the Mean

Shown below is the grouped frequency distribution of the miles-per-gallon scores of sixty 1976 automobiles. One column—Midpoint of interval—has been added.

Class interval	Midpoint of interval	f	fX
28–29	28.5	1	28.5
26–27	26.5	1	26.5
24–25	24.5	2	49.0
22–23	22.5	3	67.5
20–21	20.5	7	143.5
18–19	18.5	6	111.0
16–17	16.5	10	165.0
14–15	14.5	6	87.0
12–13	12.5	10	125.0
10–11	10.5	14	147.0
			$\Sigma fX = 950$

4. Construct a frequency polygon of the ungrouped frequency distribution of thirty miles-per-gallon scores.

Following are the steps for calculating the mean of a grouped frequency distribution.

A.   Multiply the score at the midpoint of each interval by its corresponding frequency. Record in the fX column.

B.   Sum the fX column to obtain ΣfX. This sum equals 950.

C.   Divide ΣfX by N to obtain the mean:

$$\bar{X} = \frac{\Sigma fX}{N} = \frac{950}{60} = 15.83$$

Do Exercises 5 through 7 at the right.

Combined means with equal N's

There are many occasions when we are given the means of a number of different groups and we wish to combine them in order to obtain the overall mean. *As long as the means are all based on equal N's, there is no problem.* We merely sum the means and divide by the number of means.

Example

$\bar{X}_1 = 15,$ $\bar{X}_2 = 25,$ $\bar{X}_3 = 35,$ $\bar{X}_4 = 5.$

If N's are the same, the overall mean, $\bar{X}_T$, is

$$\bar{X}_T = \frac{\sum_{i=1}^{4} \bar{X}_i}{N_M} = \frac{80}{4} = 20,$$

in which N_M is the number of means.

Do Exercises 8 and 9 at the right.

Weighted means with unequal N's

When the N's are not equal, special procedures are needed to determine the overall mean of a set of means. Recall that the mean is defined as

$$\bar{X} = \frac{\Sigma X}{N}.$$

In Chapter 1, we saw that the above may also be written $\Sigma X = N\bar{X}$.

This tells us that, if we know the mean and the N, we can obtain ΣX.

Calculate the mean for each of the following grouped frequency distributions.

5.

Class interval	Midpoint of interval	f	fX
28–29		1	
26–27		0	
24–25		1	
22–23		0	
20–21		2	
18–19		3	
16–17		5	
14–15		3	
12–13		8	
10–11		7	

6.

Class interval	Midpoint of interval	f	fX
28–30		1	
25–27		1	
22–24		5	
19–21		11	
16–18		12	
13–15		11	
10–12		19	

7.

Class interval	Midpoint of interval	f	fX
24–26		1	
21–23		2	
18–20		4	
15–17		8	
12–14		16	
9–11		8	
6–8		4	
3–5		2	
0–2		1	
		$N = 46$	

Find the overall mean of the following means. Each mean is based on the same number of observations (N).

8. $\bar{X}_1 = 5, \bar{X}_2 = 27, \bar{X}_3 = 9$

9. $\bar{X}_1 = 75.2, \bar{X}_2 = 60.4, \bar{X}_3 = 98.3,$

   $\bar{X}_4 = 84.3$

5

Examples

$\bar{X}_1 = 20,\ N_1 = 5$

$\Sigma X_1 = (20)(5)$

$\quad\ = 100$

$\bar{X}_2 = 30,\ N_2 = 8$

$\Sigma X_2 = (30)(8)$

$\quad\ = 240$

$\bar{X}_3 = 42,\ N_3 = 7$

$\Sigma X_3 = (42)(7)$

$\quad\ = 294$

Do Exercises 10 through 13 at the right.

We may now sum these ΣX's to obtain an overall sum of scores:

$\Sigma X_1 + \Sigma X_2 + \Sigma X_3 = 100 + 240 + 294$

$\qquad\qquad\qquad = 634.$

We may sum the N's to obtain an overall N_T (N total), i.e.,

$N_T = N_1 + N_2 + N_3 + \cdots + N_N.$

In the above examples, $N_1 = 5,\ N_2 = 8,\ N_3 = 7$. We find

$N_T = 20.$

The overall mean, $\bar{X}_T$, then is

$\bar{X}_T = \dfrac{634}{20} = 31.7.$

Examples

Given:

$\bar{X}_1 = 20,\ N_1 = 10;$

$\bar{X}_2 = 25,\ N_2 = \ \ 5.$

Then

$\Sigma X_1 = 200$

$\underline{\Sigma X_2 = 125}$

$\overline{\Sigma X_T = 325}$

and

$N_T = 10 + 5$

$\quad\ = 15.$

Find ΣX from which each of the following means was calculated.

10. $\bar{X}_1 = 3,\ N = 15$

11. $\bar{X}_2 = 49,\ N = 7$

12. $\bar{X}_3 = 107,\ N = 12$

13. $\bar{X}_4 = 36,\ N = 62$

Thus $\bar{X}_T = \dfrac{325}{15} = 21.67.$

Given:

$\bar{X}_1 = 85, N_1 = 20;$

$\bar{X}_2 = 81, N_2 = 27;$

$\bar{X}_3 = 75, N_3 = 55;$

$\bar{X}_4 = 90, N_4 = 10.$

Then

$\Sigma X_1 = 1700$

$\Sigma X_2 = 2187$

$\Sigma X_3 = 4125$

$\underline{\Sigma X_4 = \quad 900}$

$\Sigma X_T = 8912$

and

$N_1 = 20$

$N_2 = 27$

$N_3 = 55$

$\underline{N_4 = 10}$

$N_T = 112.$

Thus $\bar{X}_T = \dfrac{8912}{112} = 79.57.$

At various times, you obtained 10 shares of a given stock at $12 a share; 20 shares at $8 a share; 5 shares at $14 a share; 40 shares at $7 a share. Find the mean price per share.

Given:

$\bar{X}_1 = \$12, N_1 = 10;$

$\bar{X}_2 = \$\ 8, N_2 = 20;$

$\bar{X}_3 = \$14, N_3 = \ 5;$

$\bar{X}_4 = \$\ 7, N_4 = 40.$

Then

$\Sigma X_1 = \$120$

$\Sigma X_2 = \$160$

$\Sigma X_3 = \$\ 70$

$\underline{\Sigma X_4 = \$280}$

$\Sigma X_T = \$630$

and

$N_1 = 10$

$N_2 = 20$

$N_3 = \ 5$

$\underline{N_4 = 40}$

$N_T = 75.$

Thus $\quad \bar{X}_T = \dfrac{\$630}{75} = \$8.40.$

Do Exercises 14 through 17 at the right.

Calculating the Median

As indicated earlier, the median is merely a special case of a percentile rank. It is the score at the 50th percentile, and it exactly divides the distribution of scores into two halves.

The following steps are used with a grouped frequency distribution. Reproduced below is the grouped frequency distribution we employed in Chapter 4 to demonstrate the calculation of percentile ranks of scores and scores corresponding to given percentile ranks.

Real limits of interval	f	Cum f
27.5–29.5	1	60
25.5–27.5	1	59
23.5–25.5	2	58
21.5–23.5	3	56
19.5–21.5	7	53
17.5–19.5	6	46
15.5–17.5	10	40
13.5–15.5	6	30
11.5–13.5	10	24
9.5–11.5	14	14
$\Sigma f = 60$		

Find the overall mean of the following sets of means.

14. $\bar{X}_1 = \$42.5, N_1 = 6$

   $\bar{X}_2 = \$37.4, N_2 = 7$

   $\bar{X}_3 = \$52.8, N_3 = 20$

15. $\bar{X}_1 = 7, N = 15$

   $\bar{X}_2 = 20, N = 4$

16. A given stock was purchased at the following prices at various times.

   20 shares at $\$\ 8.20$ a share
   100 shares at $\$10.90$ a share
   50 shares at $\$\ 9.40$ a share
   200 shares at $\$\ 7.80$ a share

   Find the mean cost per share.

17. An automobile obtained 15 miles per gallon on 12 gallons of gas; 17 miles per gallon on 25 gallons of gas; 12 miles per gallon on 8 gallons of gas; and 18 miles per gallon on 27 gallons of gas.

   Find the overall mean miles per gallon.

A. Multiply 50 by N and divide by 100. Thus,

$$\frac{50 \times 60}{100} = 30.$$

B. Find the class interval containing the 30th frequency. This is the interval 13.5–15.5.

C. Subtract from the value obtained in step A the cumulative frequency at the upper real limit of the *adjacent lower* interval, that is,

$$30 - 24 = 6.$$

D. Divide the value found in step C by the frequency within the interval and multiply by the width of the interval ($i = 2$). Thus,

$$\frac{6}{6} \times 2 = 2.$$

E. Add the value in step D to the score at the lower real limit of the interval containing the 30th frequency. Thus,

$$13.5 + 2.0 = 15.5$$

is the answer.

Do Exercises 18 through 20 at the right.

Obtaining the Mode

The mode is the midpoint of the interval with the highest associated frequency. In the distribution of miles per gallon of sixty automobiles, the interval 10–11 has the highest associated frequency (14). The midpoint of this interval is 10.5; therefore, the mode is 10.5

Do Exercises 21 through 24 at the right.

Calculate the median in each of the following grouped frequency distributions.

18.

Real limits of interval	f	Cum f
27.5–30.5	1	
24.5–27.5	1	
21.5–24.5	5	
18.5–21.5	11	
15.5–18.5	12	
12.5–15.5	11	
9.5–12.5	19	

19.

Real limits of interval	f	Cum f
27.5–29.5	1	
25.5–27.5	0	
23.5–25.5	1	
21.5–23.5	0	
19.5–21.5	2	
17.5–19.5	3	
15.5–17.5	5	
13.5–15.5	3	
11.5–13.5	8	
9.5–11.5	7	

20. The grouped frequency distribution in Exercise 7.

Find the mode of the following:

21. The distribution in Exercise 5.

22. The distribution in Exercise 6.

23. The distribution in Exercise 7.

24. The distribution in Exercise 7 is symmetrical, i.e., there are identical frequency counts for corresponding intervals on either side of the middle interval. Note the mean, median, and mode you obtained in Exercises 7, 20, and 23. Form a generalization:

In a symmetrical distribution with a single mode, the mean, median, and mode are _____.

CHAPTER 5 TEST

Problems 1 and 2 are based on the following set of scores.

15	13
14	13
14	12
13	12
13	11

1. The mean of these scores is:

   a) 12.5   b) 13.5   c) 13.0   d) 12.9

2. The mode of these scores is:

   a) 13   b) 4   c) 12.5   d) 13.5

Problems 3 through 5 are based on the following ungrouped frequency distribution.

X	f
15	1
14	6
13	12
12	8
11	2

3. The mean is:

   a) 12.43   b) 35.00   c) 6.03   d) 12.86

4. The mode is:

   a) 12   b) 13   c) 13.5   d) 12.5

5. The median is:

   a) 12.8   b) 12.9   c) 13.2   d) 14.9

Problems 6 through 8 are based on the following grouped frequency distribution.

Class interval	f
22–24	2
19–21	3
16–18	5
13–15	8
10–12	4
7– 9	1

6. The mode is:

   a) 8   b) 13   c) 14   d) 15

7. The median is:

   a) 14.9   b) 13.3   c) 17.3   d) 13.9

8. The mean is:

   a) 17.95   b) 16.14   c) 6.48   d) 15.43

Problems 9 through 11 are based on the following grouped frequency distribution.

Class interval	f
56–64	2
47–55	5
38–46	12
29–37	15
20–28	13
11–19	6
2–10	1

9. The mode is:

    a) 15   b) 29   c) 37   d) 33

10. The median is:

    a) 32.2   b) 32.7   c) 42.2   d) 33.0

11. The mean is:

    a) 34.15   b) 32.86   c) 29.45   d) 33.00

12. If $\bar{X}_1 = 2.53$, $\bar{X}_2 = 1.97$, $\bar{X}_3 = 4.05$, $\bar{X}_4 = 3.33$ and all are based on N's of 8, $\bar{X}_T$ equals:

    a) 11.88   b) 1.48   c) 0.37   d) 2.97

13. If you purchased 8 shares of stock at $9.30 a share, 22 shares at $25.42 a share, and 40 shares of stock at $26.89 a share, the total purchase price for all shares was:

    a) $347.36   b) $26.28   c) $1839.44   d) $1075.60

14. If you purchase 6 pounds of beef at one supermarket at $1.38 per pound, 15 pounds at another market at $1.28 per pound, and 9 pounds at a third market at $1.54 per pound, what is the mean price per pound?

    a) $1.38   b) $13.78   c) $41.34   d) $1.40

15. Find the overall mean of the following set of means:

    $\bar{X}_1 = 16.4$, $N_1 = 15$

    $\bar{X}_2 = 26.3$, $N_2 = 16$

    $\bar{X}_3 = 14.1$, $N_3 = 14$

    a) 57.61   b) 19.20   c) 18.93   d) 15.00

Practical

Problems 16 and 17 are based on the following set of scores.

15	21	21	44	15	23
31	25	8	34	27	33
23	23	41	24	23	5
31	41	20	17	18	18
29	39	24	38	22	25
22	12	28	7	35	28
15	40	24	2	26	23
23	25	22	13	6	5

16. Group the given scores into a grouped frequency distribution, $i = 4$.

17. Calculate the mean, median, and mode of the grouped frequency distribution derived in Problem 1.

18. Find the weighted mean for the following sets of means.

a) $\bar{X}_1 = 8.7, N_1 = 15$

   $\bar{X}_2 = 12.5, N_2 = 6$

   $\bar{X}_3 = 11.2, N_3 = 11$

   $\bar{X}_4 = 9.8, N_4 = 14$

b) $\bar{X}_1 = 85, N_1 = 30$

   $\bar{X}_2 = 70, N_2 = 28$

   $\bar{X}_3 = 63, N_3 = 34$

   $\bar{X}_4 = 77, N_4 = 36$

   $\bar{X}_5 = 92, N_5 = 28$

   $\bar{X}_6 = 68, N_6 = 23$

c) $\bar{X}_1 = 18, N_1 = 15$

   $\bar{X}_2 = 14, N_2 = 12$

   $\bar{X}_3 = 7, N_3 = 22$

   $\bar{X}_4 = 12, N_4 = 11$

Measures of Dispersion

MEASURES OF DISPERSION

Just as measures of central tendency generally tell us something about the central portions of a set of scores, measures of dispersion tell us something about the degree of spread, variability, or dispersion of these scores. Each of the following have the same mean, median, and mode but they differ in dispersion.

Examples

The set of numbers 38, 40, 40, 42 is less dispersed than the set 30, 40, 40, 50.

The set of numbers 10, 90, 90, 170 is more dispersed or more variable than the set 85, 90, 90, 95.

The set of numbers 5, 6, 6, 7 is less variable than the set 4, 6, 6, 8.

Do Exercises 1 through 5 at the right.

THE CRUDE RANGE

The crude range is the most straightforward measure of variability. It is merely the largest number in a set of scores minus the lowest.

Examples

The range of the following scores—38, 40, 40, 42—is

42 − 38 = 4.

The range of the following quantities—30, 40, 40, 50—is

50 − 30 = 20.

The range of the sixty miles-per-gallon scores is

29 − 10 = 19.

The range of the thirty miles-per-gallon scores is

28 − 10 = 18.

Do Exercises 6 through 12 at the right.

OBJECTIVES

- Understand the concept of variability or dispersion.
- Know how to calculate the crude range.
- Know how to calculate the interquartile range.
- Know how to calculate the mean deviation.
- Know how to calculate the variance and standard deviation using:
  a) the mean-deviation method,
  b) raw-score method applied to a set of scores,
  c) raw-score method from an ungrouped frequency distribution,
  d) raw-score method from a grouped frequency distribution.

EXERCISES

Indicate which of the paired sets of numbers shows greater variability.

1. a) 100, 150, 200
   b) 50, 150, 250

2. a) 12, 15, 18
   b) 13, 15, 17

3. a) 7, 19, 35
   b) 97, 99, 100

4. a) 0, 5, 10
   b) 60, 62, 64

5. a) 2, 8, 14
   b) 90, 110, 130

Determine the crude range in each of the following sets of scores.

6. 2, 5, 9, 26

7. 105, 146, 158, 183

8. 109, 102, 100, 92

9. 1465, 984, 672, 101

10. 15, 83, 5, 109, 75

11. 17, 29, 44, 15, 3

12. 84, 85, 86, 87, 405

INTERQUARTILE RANGE

The crude range is the victim of the two most extreme scores. If one or both of these scores is not typical, the crude range will provide a distorted view of variability (see Exercise 12 at right as an example).

The interquartile range is a more stable measure of variability. It consists of specifying the score at both the 25th percentile and the 75th percentile. Fifty percent of the scores will fall within the interquartile range. Also, twenty-five percent of the scores will fall below the lower quartile and another twenty-five percent above the 75th percentile (or third quartile).

Example

In the grouped frequency distribution ($i = 2$) of the mpg of 60 cars, the median is a score of 15.5. The 25th percentile is a score of 11.70; the score at the 75th percentile is 19.17. The interquartile range is 11.70–19.17.

Do Exercises 13 and 14 at the right.

THE MEAN DEVIATION

Look at the following two sets of scores:

Set A: 3, 5, 7
Set B: 1, 5, 9

Both have the same mean, but Set B is clearly more variable than Set A. Another method to denote this difference in variability would be to express each score as a deviation from the mean and then find the mean of these deviations.

Examples

Set A		Set B	
X	$X - \bar{X}$	X	$X - \bar{X}$
3	−2	1	−4
5	0	5	0
7	+2	9	+4
$\bar{X}_A = 5$	$\Sigma(X - \bar{X}_A) = 0$	$\bar{X}_B = 5$	$\Sigma(X - \bar{X}_B) = 0$

The problem is that the sum of the deviations of scores about the mean is zero.

Do Exercises 15 and 16 at the right.

Calculate the interquartile range of the following grouped frequency distributions.

13.

Real limits of interval	f	Cum f
27.5–30.5	1	
24.5–27.5	1	
21.5–24.5	5	
18.5–21.5	11	
15.5–18.5	12	
12.5–15.5	11	
9.5–12.5	19	

14.

Real limits of interval	f	Cum f
27.5–29.5	1	
25.5–27.5	0	
23.5–25.5	1	
21.5–23.5	0	
19.5–21.5	2	
17.5–19.5	3	
15.5–17.5	5	
13.5–15.5	3	
11.5–13.5	8	
9.5–11.5	7	

Confirm that the sum of the deviations of each score from the mean is zero.

15.

X	$(X - \bar{X})$
105	
110	
115	

16.

X	$(X - \bar{X})$
20	
25	
25	
30	
30	
30	
35	
35	
40	

We could overcome this problem by adding the deviations *without regard to sign* and then dividing by N. The resulting measure, the mean deviation, would reflect the extent of deviations of scores from the mean. The resulting statistic would be based on the *absolute value* of the deviations. The absolute value of a positive number or zero is the number itself. The absolute value of a negative number can be found by changing the sign from negative to positive. Thus, the absolute value of $+8$ or -8 is 8. The symbol for an absolute value is $|\ |$. Thus, $|-8| = 8$.

Do Exercises 17 through 23 at the right.

To find the mean deviation:

A. Calculate the mean of a set of scores.
B. Subtract the mean from each score $(X - \bar{X})$. The resulting value is referred to as the deviation from the mean.
C. Sum the deviations without regard to sign.
D. Divide by N.

Examples

Set A		Set B	
X	$|X - \bar{X}|$	X	$|X - \bar{X}|$
3	$|-2|$	1	$|-4|$
5	$|0|$	5	$|0|$
7	$|+2|$	9	$|+4|$
$\bar{X}_A = 5$	$\Sigma(|X - \bar{X}|) = 4$	$\bar{X}_B = 5$	$\Sigma(|X - \bar{X}|) = 8$

$$\text{M.D.} = \frac{4}{3} = 1.33 \qquad\qquad \text{M.D.} = \frac{8}{3} = 2.67$$

Do Exercises 24 through 26 at the right.

Express the following numbers as absolute values.

17. -24

18. 15

19. -1

20. 0

21. -453

22. 1

23. -0.06

Find the mean deviation of the following sets of scores.

24.

X	$|X - \bar{X}|$
115	
110	
105	

25.

X	$|X - \bar{X}|$
40	
35	
35	
30	
30	
30	
25	
25	
20	

26.

X	$|X - \bar{X}|$
15	
13	
12	
10	
8	
7	
5	

6

The procedures with grouped and ungrouped frequency distributions are the same except that, with grouped frequency distributions, you multiply each deviation score by its corresponding frequency prior to summing.

Examples

X	f	$|X - \bar{X}|$	$f(|X - \bar{X}|)$
15	2	$|4|$	8
14	2	$|3|$	6
13	4	$|2|$	8
12	6	$|1|$	6
11	9	$|0|$	0
10	7	$|-1|$	7
9	1	$|-2|$	2
8	5	$|-3|$	15
7	1	$|-4|$	4
$\bar{X} = 11$ $N = 37$			$\Sigma f(|X - \bar{X}|) = 56$

M.D. $= \dfrac{\Sigma f(|X - \bar{X}|)}{N} = \dfrac{56}{37} = 1.51$

Do Exercises 27 and 28 at the right.

THE VARIANCE AND THE STANDARD DEVIATION

The variance and standard deviation are the most useful and widely employed measures of variability or dispersion. The standard deviation permits us to directly compare the variability of different sets of scores, allows the precise interpretation of scores within a set, and is a member of a mathematical system which permits its use in more advanced statistical considerations.

The variance is defined as the sum of the squared deviations about the mean divided by $N - 1$.

$$s^2 = \frac{\Sigma(X - \bar{X})^2*}{N - 1}$$

The standard deviation is, in turn, the square root of the variance, i.e., $s = \sqrt{s^2}$. The calculation of both requires but the simple extension of the procedure for calculating the mean deviation.

---

* In the descriptive sections of many text books the variance s^2 is defined by the use of N instead of $N - 1$ in the denominator. If this is the case with your text, skip this section and go on to the next (p. 74).

Calculate the mean deviation for the following frequency distributions.

27.

X	f	$|X - \bar{X}|$	$f(|X - \bar{X}|)$
40	1		
35	2		
30	3		
25	2		
20	1		
$\bar{X} = 30$			

28.

Class interval	Mid-point of interval	f	$|X - \bar{X}|$	$f(|X - \bar{X}|)$
16–18	17	2		
13–15	14	5		
10–12	11	9		
7– 9	8	9		
4– 6	5	5		
1– 3	2	2		
$\bar{X} = 9.5$				

Calculating s^2 and s from a Set of Scores—Mean-Deviation Method

The following steps are involved in calculating the variance and standard deviation from a set of scores.

A. Find the mean of the scores.
B. Subtract the mean from each score and then square.
C. Sum all of the squared deviations.
D. Substitute in the formulas for s^2 and s.

Examples

X	$X - \bar{X}$	$(X - \bar{X})^2$
15	5	25
13	3	9
12	2	4
10	0	0
8	-2	4
7	-3	9
5	-5	25
$\Sigma X = 70$	$\Sigma(X - \bar{X}) = 0$	$\Sigma(X - \bar{X})^2 = 76$

$$\bar{X} = \frac{70}{7} = 10$$

$$s^2 = \frac{\Sigma(X - \bar{X})^2}{N - 1} = \frac{76}{6} = 12.67$$

$$s = \sqrt{s^2} = \sqrt{12.67} = 3.56$$

Do Exercises 29 through 32 at the right.

Calculating s^2 and s from a Set of Scores—Raw-Score Method

It is often inconvenient to calculate s^2 and s by use of the mean-deviation procedures. This is particularly the case when the mean is a fractional value. The raw-score method is also extremely useful when automatic calculators are available.

The term $\Sigma(X - \bar{X})^2$ is referred to as the *sum of squares*. The raw-score formula for calculating the sum of squares is

$$\Sigma(X - \bar{X})^2 = \Sigma X^2 - \frac{(\Sigma X)^2}{N}.$$

Find the variance and standard deviations of the following sets of scores.

29.

X	$X - \bar{X}$	$(X - \bar{X})^2$
105		
110		
115		
120		
$\Sigma X =$		
$\bar{X} =$		
$s^2 =$		
$s =$		

30.

X	$X - \bar{X}$	$(X - \bar{X})^2$
5		
10		
15		
20		
$\Sigma X =$		
$\bar{X} =$		
$s^2 =$		
$s =$		

31.

X	$X - \bar{X}$	$(X - \bar{X})^2$
27		
32		
37		
42		
$\Sigma X =$		
$\bar{X} =$		
$s^2 =$		
$s =$		

32.

X	$X - \bar{X}$	$(X - \bar{X})^2$
1		
8		
15		
16		
$\Sigma X =$		
$\bar{X} =$		
$s^2 =$		
$s =$		

6

The following steps are employed in calculating s^2 and s from a set of scores using the raw-score method.

X	X^2
15	225
13	169
12	144
10	100
8	64
7	49
5	25
$\Sigma X = 70$	$\Sigma X^2 = 776$

$N = 7$

A. Find ΣX. In the present example $\Sigma X = 70$. Square this value to obtain $(\Sigma X)^2 = 4900$.

B. Square each X and place in the column headed by X^2.

C. Sum all the squared scores. In the present example $\Sigma X^2 = 776$.

D. Substitute ΣX^2, $(\Sigma X)^2$, and N in the appropriate place in the formula for the sum of squares.

$$\Sigma(X - \bar{X})^2 = \Sigma X^2 - \frac{(\Sigma X)^2}{N}$$

$$= 776 - \frac{4900}{7}$$

$$= 776 - 700$$

$$= 76$$

E. Substitute this value in the formula for s^2.

$$s^2 = \frac{\Sigma(X - \bar{X})^2}{N - 1}$$

$$= \frac{76}{6} = 12.67$$

F. The standard deviation, s, is the square root of s^2.

$$s = \sqrt{s^2}$$

$$= \sqrt{12.67}$$

$$= 3.56$$

Do Exercises 33 through 35 at the right.

Find s^2 and s, using the raw-score method.

33.

X	X^2
20	
15	
10	
5	
$\Sigma X =$	$\Sigma X^2 =$

$$\Sigma(X - \bar{X})^2 = \Sigma X^2 - \frac{(\Sigma X)^2}{N}$$

$$s^2 = \frac{\Sigma(X - \bar{X})^2}{N - 1}$$

34.

X	X^2
23	
17	
15	
14	
11	
9	
5	

35.

X	X^2
16.5	
13.2	
11.6	
10.4	
9.3	
7.1	
4.5	

Calculating s and s^2 from an Ungrouped Frequency Distribution—Raw-Score Method

The procedures for calculating s and s^2 from an ungrouped frequency distribution are the same as the calculation of those measures of variability from a set of scores except that each squared score must be multiplied by its corresponding frequency prior to summing.

The sum of squares is defined as

$$\Sigma f(X - \bar{X})^2 = \Sigma fX^2 - \frac{(\Sigma fX)^2}{N}.$$

And,

$$s^2 = \frac{\Sigma f(X - \bar{X})^2}{N - 1},$$

$$s = \sqrt{s^2}.$$

Example

X	f	fX	X^2	fX^2
40	1	40	1600	1600
35	2	70	1225	2450
30	3	90	900	2700
25	2	50	625	1250
20	1	20	400	400
$N = 9$		$\Sigma fX = 270$		$\Sigma fX^2 = 8400$

A. Multiply each score by its corresponding frequency to obtain fX.

B. Sum the fX column to obtain ΣfX. In the present example $\Sigma fX = 270$. Square this value to obtain $(\Sigma fX)^2 = 72{,}900$.

C. Square each score and place in X^2 column.

D. Multiply each X^2 by its corresponding frequency and place in the fX^2 column.

E. Sum the values in the fX^2 column to obtain ΣfX^2. In the present example

$$\Sigma fX^2 = 8400.$$

F. Substitute ΣfX^2, $(\Sigma fX)^2$, and N in the formula for $\Sigma f(X - \bar{X})^2$. In the present example

$$\Sigma f(X - \bar{X})^2 = 8400 - \frac{72{,}900}{9}$$

$$= 8400 - 8100$$

$$= 300.$$

G. Substitute $\Sigma f(X - \bar{X})^2$ in the formula for s^2 and solve.

$$s^2 = \frac{\Sigma f(X - \bar{X})^2}{N - 1}$$

$$= \frac{300}{8}$$

$$= 37.5$$

H. To obtain s, find the square root of s^2.

$$s = \sqrt{s^2}$$

$$= \sqrt{37.5}$$

$$= 6.12$$

Do Exercise 36 at the right.

Calculating s^2 and s From a Grouped Frequency Distribution—Raw-Score Method

The procedures for calculating s and s^2 from a grouped frequency distribution are the same as the calculation of these measures of variability from an ungrouped frequency distribution except that midpoints of intervals must be used in the calculations.

Example

Class interval	Midpoint of interval	f	fX	X^2	fX^2
28–29	28.5	1	28.5	812.25	812.25
26–27	26.5	1	26.5	702.25	702.25
24–25	24.5	2	49.0	600.25	1200.50
22–23	22.5	3	67.5	506.25	1518.75
20–21	20.5	7	143.5	420.25	2941.75
18–19	18.5	6	111.0	342.25	2053.50
16–17	16.5	10	165.0	272.25	2722.50
14–15	14.5	6	87.0	210.25	1261.50
12–13	12.5	10	125.0	156.25	1562.50
10–11	10.5	14	147.0	110.25	1543.50
		$\Sigma f = 60$	$\Sigma fX = 950.0$		$\Sigma fX^2 = 16{,}319.00$

A. Multiply the midpoint of each interval by its corresponding frequency to obtain fX.

B. Sum the fX column to obtain ΣfX. In the present example $\Sigma fX = 950$. Square this value to obtain $(\Sigma fX)^2 = 902{,}500$.

36. Find the standard deviation of the following frequency distribution.

X	f
15	1
14	4
13	8
12	16
11	12
10	2
9	1

C. Square the midpoint of each interval and place in X^2 column.

D. Multiply each X^2 by its corresponding frequency and place in the fX^2 column.

E. Sum the values in the fX^2 column to obtain ΣfX^2. In the present example

$\Sigma fX^2 = 16{,}319.$

F. Substitute ΣfX^2, $(\Sigma fX)^2$, and N in the formula for the sum of squares, i.e., $\Sigma f(X - \bar{X})^2$. In the present example

$$\Sigma f(X - \bar{X})^2 = 16{,}319 - \frac{902{,}500}{60}$$

$$= 16{,}319 - 15{,}041.67$$

$$= 1277.33.$$

G. Substitute $\Sigma f(X - \bar{X})^2$ in the formula for s^2 and solve.

$$s^2 = \frac{\Sigma f(X - \bar{X})^2}{N - 1}$$

$$= \frac{1277.33}{59}$$

$$= 21.65$$

H. To obtain s, find the square root of s^2.

$$s = \sqrt{s^2}$$

$$= \sqrt{21.65}$$

$$= 4.65$$

Do Exercises 37 and 38 at the right.

THE VARIANCE AND STANDARD DEVIATION: ALTERNATE SECTION

This section is employed to accompany those texts in which the variance and standard deviation are defined by the use of N, instead of $N - 1$, in the denominator.

The variance and standard deviation are the most useful and widely employed measures of variability or dispersion. The standard deviation permits us to directly compare the variability of different sets of scores, allows the precise interpretation of scores within a set, and is a member of mathematical system which permits its use in more advanced statistical considerations.

Find s^2 and s for the following grouped frequency distributions.

37.

Class interval	Midpoint of interval	f	fX	X^2	fX^2
28–30		1			
25–27		1			
22–24		5			
19–21		11			
16–18		12			
13–15		11			
10–12		19			

38.

Class interval	Midpoint of interval	f	fX	X^2	fX^2
28–29		1			
26–27		0			
24–25		1			
22–23		0			
20–21		2			
18–19		3			
16–17		5			
14–15		3			
12–13		8			
10–11		7			

6

The variance is defined as the sum of the squared deviations about the mean divided by N:

$$s^2 = \frac{\Sigma(X - \bar{X})^2}{N}$$

The standard deviation is, in turn, the square root of the variance, i.e., $s = \sqrt{s^2}$. The calculation of both requires but the simple extension of the procedure for calculating the mean deviation.

Calculating s^2 and s from a Set of Scores—Mean-Deviation Method

The following steps are involved in calculating the variance and standard deviation from a set of scores.

A. Find the mean of the scores.

B. Subtract the mean from each score and then square.

C. Sum all of the squared deviations.

D. Substitute in the formulas for s^2 and s.

Example

X	$X - \bar{X}$	$(X - \bar{X})^2$
15	5	25
13	3	9
12	2	4
10	0	0
8	−2	4
7	−3	9
5	−5	25
$\Sigma X = 70$	$\Sigma(X - \bar{X}) = 0$	$\Sigma(X - \bar{X})^2 = 76$

$$\bar{X} = \frac{70}{7} = 10$$

$$s^2 = \frac{\Sigma(X - \bar{X})^2}{N} = \frac{76}{7} = 10.86$$

$$s = \sqrt{s^2} = \sqrt{10.86} = 3.30$$

Do Exercises 29A through 32A at the right.

Calculating s^2 and s from a Set of Scores—Raw-Score Method

It is often inconvenient to calculate s^2 and s by use of the mean-deviation procedures. This is particularly the case when the mean is a fractional value. The raw-score method is also extremely useful when automatic calculators are available.

Find the variance and standard deviations of the following sets of scores.

29A.

X	$X - \bar{X}$	$(X - \bar{X})^2$
105		
110		
115		
120		
$\Sigma X =$		
$\bar{X} =$		
$s^2 =$		
$s =$		

30A.

X	$X - \bar{X}$	$(X - \bar{X})^2$
5		
10		
15		
20		
$\Sigma X =$		
$\bar{X} =$		
$s^2 =$		
$s =$		

31A.

X	$X - \bar{X}$	$(X - \bar{X})^2$
27		
32		
37		
42		
$\Sigma X =$		
$\bar{X} =$		
$s^2 =$		
$s =$		

32A.

X	$X - \bar{X}$	$(X - \bar{X})^2$
1		
8		
15		
16		
$\Sigma X =$		
$\bar{X} =$		
$s^2 =$		
$s =$		

The term $\Sigma(X - \bar{X})^2$ is referred to as the *sum of squares*. The raw-score formula for calculating the sum of squares is

$$\Sigma(X - \bar{X})^2 = \Sigma X^2 - \frac{(\Sigma X)^2}{N}.$$

The following steps are employed in calculating s^2 and s from a set of scores using the raw-score method.

X	X^2
15	225
13	169
12	144
10	100
8	64
7	49
5	25
$\Sigma X = 70$	$\Sigma X^2 = 776$

$N = 7$

A. Find ΣX. In the present example $\Sigma X = 70$. Square this value to obtain $(\Sigma X)^2 = 4900$.

B. Square each X and place in the column headed by X^2.

C. Sum all the squared scores. In the present example $\Sigma X^2 = 776$.

D. Substitute ΣX^2, $(\Sigma X)^2$, and N in the appropriate place in the formula for the sum of squares.

$$\Sigma(X - \bar{X})^2 = \Sigma X^2 - \frac{(\Sigma X)^2}{N}$$

$$= 776 - \frac{4900}{7}$$

$$= 776 - 700$$

$$= 76$$

E. Substitute this value in the formula for s^2.

$$s^2 = \frac{\Sigma(X - \bar{X})^2}{N}$$

$$= \frac{76}{7} = 10.86$$

F. The standard deviation, s, is the square root of s^2.

$$s = \sqrt{s^2}$$

$$= \sqrt{10.86}$$

$$= 3.30$$

Do Exercises 33A through 35A at the right.

Find s^2 and s, using the raw-score method.

33A.

X	X^2
20	
15	
10	
5	
$\Sigma X =$	$\Sigma X^2 =$

$$\Sigma(X - \bar{X})^2 = \Sigma X^2 - \frac{(\Sigma X)^2}{N}$$

$$s^2 = \frac{\Sigma(X - \bar{X})^2}{N - 1}$$

34A.

X	ΣX^2
23	
17	
15	
14	
11	
9	
5	

35A.

X	ΣX^2
16.5	
13.2	
11.6	
10.4	
9.3	
7.1	
4.5	

Calculating s and s^2 from an Ungrouped Frequency Distribution—Raw-Score Method

The procedures for calculating s and s^2 from an ungrouped frequency distribution are the same as the calculation of those measures of variability from a set of scores except that each squared score must be multiplied by its corresponding frequency prior to summing.

The sum of squares is defined as

$$\Sigma f(X - \bar{X})^2 = \Sigma fX^2 - \frac{(\Sigma fX)^2}{N}.$$

And,

$$s^2 = \frac{\Sigma f(X - \bar{X})^2}{N},$$

$$s = \sqrt{s^2}.$$

Example

X	f	fX	X^2	fX^2
40	1	40	1600	1600
35	2	70	1225	2450
30	3	90	900	2700
25	2	50	625	1250
20	1	20	400	400
	$N = 9$	$\Sigma fX = 270$		$\Sigma fX^2 = 8400$

A. Multiply each score by its corresponding frequency to obtain fX.

B. Sum the fX column to obtain ΣfX. In the present example $\Sigma fX = 270$. Square this value to obtain $(\Sigma fX)^2 = 72{,}900$.

C. Square each score and place in X^2 column.

D. Multiply each X^2 by its corresponding frequency and place in the fX^2 column.

E. Sum the values in the fX^2 column to obtain ΣfX^2. In the present example

$$\Sigma fX^2 = 8400.$$

F. Substitute ΣfX^2, $(\Sigma fX)^2$, and N in the formula for $\Sigma f(X - \bar{X})^2$. In the present example

$$\Sigma f(X - \bar{X})^2 = 8400 - \frac{72{,}900}{9}$$

$$= 8400 - 8100$$

$$= 300.$$

G. Substitute $\Sigma f(X - \bar{X})^2$ in the formula for s^2 and solve.

$$s^2 = \frac{\Sigma f(X - \bar{X})^2}{N}$$

$$= \frac{300}{9}$$

$$= 33.33$$

H. To obtain s, find the square root of s^2.

$$s = \sqrt{s^2}$$

$$= \sqrt{33.33}$$

$$= 5.77$$

Do Exercise 36A at the right.

Calculating s^2 and s From a Grouped Frequency Distribution—Raw-Score Method

The procedures for calculating s and s^2 from a grouped frequency distribution are the same as the calculation of these measures of variability from an ungrouped frequency distribution except that midpoints of intervals must be used in the calculations.

Example

Class interval	Midpoint of interval	f	fX	X^2	fX^2
28–29	28.5	1	28.5	812.25	812.25
26–27	26.5	1	26.5	702.25	702.25
24–25	24.5	2	49.0	600.25	1200.50
22–23	22.5	3	67.5	506.25	1518.75
20–21	20.5	7	143.5	420.25	2941.75
18–19	18.5	6	111.0	342.25	2053.50
16–17	16.5	10	165.0	272.25	2722.50
14–15	14.5	6	87.0	210.25	1261.50
12–13	12.5	10	125.0	156.25	1562.50
10–11	10.5	14	147.0	110.25	1543.50
		$\Sigma f = 60$	$\Sigma fX = 950.0$		$\Sigma fX^2 = 16{,}319.00$

A. Multiply the midpoint of each interval by its corresponding frequency to obtain fX.

B. Sum the fX column to obtain ΣfX. In the present example $\Sigma fX = 950$. Square this value to obtain $(\Sigma fX)^2 = 902{,}500$.

36A. Find the standard deviation of the following frequency distribution.

X	f
15	1
14	4
13	8
12	16
11	12
10	2
9	1

C. Square the midpoint of each interval and place in X^2 column.

D. Multiply each X^2 by its corresponding frequency and place in the fX^2 column.

E. Sum the values in the fX^2 column to obtain ΣfX^2. In the present example

$\Sigma fX^2 = 16{,}319.$

F. Substitute ΣfX^2, $(\Sigma fX)^2$, and N in the formula for the sum of squares, i.e., $\Sigma f(X - \bar{X})^2$. In the present example

$$\Sigma f(X - \bar{X})^2 = 16{,}319 - \frac{902{,}500}{60}$$

$$= 16{,}319 - 15{,}041.67$$

$$= 1277.33.$$

G. Substitute $\Sigma f(X - \bar{X})^2$ in the formula for s^2 and solve.

$$s^2 = \frac{\Sigma f(X - \bar{X})^2}{N}$$

$$= \frac{1277.33}{60}$$

$$= 21.29$$

H. To obtain s, find the square root of s^2.

$$s = \sqrt{s^2}$$

$$= \sqrt{21.29}$$

$$= 4.61$$

Do Exercises 37A and 38A at the right.

Find s^2 and s for the following grouped frequency distributions.

37A.

Class interval	Midpoint of interval	f	fX	X^2	fX^2
28–30		1			
25–27		1			
22–24		5			
19–21		11			
16–18		12			
13–15		11			
10–12		19			

38A.

Class interval	Midpoint of interval	f	fX	X^2	fX^2
28–29		1			
26–27		0			
24–25		1			
22–23		0			
20–21		2			
18–19		3			
16–17		5			
14–15		3			
12–13		8			
10–11		7			

CHAPTER 6 TEST

1. Following are four sets of measures. Which set shows the least variability?

   Set 1: 9, 10, 20

   Set 2: 0, 5, 9

   Set 3: 50, 60, 70

   Set 4: 31, 33, 35

   a) Set 1   b) Set 2   c) Set 3   d) Set 4

2. In the preceding problem, which set shows the greatest variability?

   a) Set 1   b) Set 2   c) Set 3   d) Set 4

3. What is the range of the following scores?

   29, 3, 145, 17, 89

   a) 60   b) 142   c) 89   d) 145

4. The scores at various percentile ranks were:

   3rd, $X = 10$ 16th, $X = 75$ 25th, $X = 80$

   30th, $X = 85$ 50th, $X = 100$ 70th, $X = 115$

   75th, $X = 120$ 84th, $X = 125$ 97th, $X = 130$

   The interquartile range is:

   a) 70–130   b) 25–75   c) 16–120   d) 80–120

5. The mean deviation of the scores 12, 15, 18 is:

   a) 2   b) 3   c) 0   d) 6

6. The standard deviation of the scores 5, 8, 13, 16 is:

   a) 4.93   b) 514   c) 4.27   d) 18.25

7. The variance of the scores 0, 5, 10, 15 is:

   a) 6.45   b) 31.25   c) 41.67   d) 116.67

8. The sum of squares of the scores 25, 28, 31, 34, 37 is:

   a) 4895   b) 90   c) 24025   d) 155

9. The sum of squares of the scores 8, 11, 11, 11, 14 is:

   a) 18   b) 25   c) 623   d) 263

10. The sum of squares of the scores 0, 2, 2, 2, 4 is:

    a) 10   b) 28   c) 20   d) 8

11. The standard deviation of the scores 6, 11, 16 is:

    a) 4.08   b) 33   c) 5.00   d) 137.67

12. The standard deviation of the scores 30, 35, 40 is:

    a) 25   b) 5.00   c) 35.00   d) 408

6

13. The variance of the scores 8, 12, 14, 15 is:

    a) 9.58   b) 16.33   c) 12.25   d) 7.19

14. The variance of the scores 0, 8, 9, 14 is:

    a) 25.19   b) 33.58   c) 5.80   d) 5.02

15. The variance of the scores 20, 23, 26, 29, 32 is:

    a) 7.2   b) 18   c) 51.84   d) 22.50

Practical

16. Find $\bar{X}$, s^2, and s for the following ungrouped frequency distribution.

X	f
23	1
22	0
20	5
19	17
18	15
17	6
16	3
15	2

17. Find $\bar{X}$, s^2, and s for the following grouped frequency distribution.

Class interval	f
44–47	1
40–43	2
36–39	2
32–35	4
28–31	4
24–27	5
20–23	8
16–19	5
12–15	5
8–11	4

The Standard Deviation and the Standard Normal Distribution

A score in and of itself is meaningless. In the past two chapters we have learned how to calculate two broad classes of statistics—Measures of Central Tendency and Dispersion—that provide a frame of reference for interpreting scores. In this chapter we shall show how the mean and standard deviation can be used to interpret scores of normally distributed variables.

THE STANDARD NORMAL DISTRIBUTION

In this chapter we shall assume that the variables with which we deal are normally distributed. The graphic form of a normal distribution is a bell-shaped curve which is symmetrical about the three measures of central tendency. In a normal distribution, the mean, median, and mode are identical values.

The graph below shows a normal distribution.

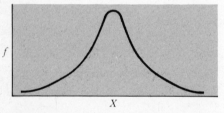

The following two figures are symmetrical but not bell-shaped. The mean and median are identical values but the mode has two values, both of which are different from the mean and median.

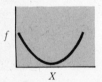   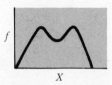

The following graphs show two bell-shaped distributions but the figure on the left is negatively skewed and the figure on the right is positively skewed.

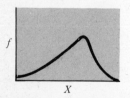   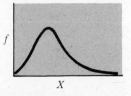

Do Exercises 1 through 5 at the right.

OBJECTIVES

- Learn how to calculate a standard score.
- Know how to interpret a score in relation to the standard normal distribution.
- Know how to compare two or more scores on a single variable.
- Know how to compare scores from two or more normally distributed variables.
- Know how to calculate the coefficient of variation.

EXERCISES

Check which of the following appear to be graphs of normal distribution.

1.
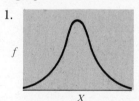

2.
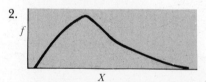

3.
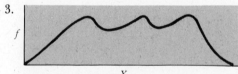

4.
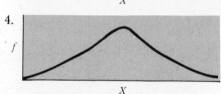

5.
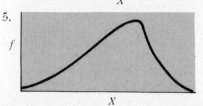

The standard normal distribution has a mean equal to zero and a standard deviation of one. It has many desirable characteristics that make it extremely useful in statistical analysis. For example, there is a fixed proportion of area between a vertical line, or ordinate, erected at any point and an ordinate erected at any other point.

The figure below shows the areas between selected points under the normal curve.

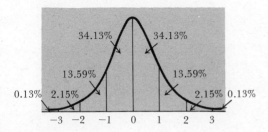

Examples

Under a normal curve, 34.13% of the area falls between 0 and 1 standard deviation above the mean.

Between the mean (0) and 2 standard deviations below the mean falls 47.72% (13.59 + 34.13) of the area.

Beyond 1 standard deviation above the mean falls 15.87% (13.59 + 2.15 + 0.13) of the area.

Always, 50% (34.13 + 13.59 + 2.15 + 0.13) of the area falls above the mean.

Do Exercises 6 through 12 at the right.

Converting Normally Distributed Variables to Units of the Standard Normal Curve

Most normally distributed variables have means and standard deviations other than 0 and 1, respectively. To employ the standard normal curve to interpret values of these variables, it is necessary to convert these scores to units of the standard normal curve. The procedures involve the transformation of a score to a z-score, also called a standard score.

Two steps are involved in this transformation: (1) subtracting the mean from the score, and (2) dividing by the standard deviation of the variable.

State the percentage of area falling between the indicated points of the standard normal curve.

6. Between the mean and 1 standard deviation below the mean.

7. Between +1 and +2 standard deviations.

8. Below the mean.

9. Between −1 and +3 standard deviations.

10. Between −3 and −1 standard deviations.

11. Between +1 and +3 standard deviations.

12. Between the mean and 2 standard deviations above the mean.

Examples

The mean of a normally distributed variable is 100 and the standard deviation is 10. Convert the following to z-scores.

Given score: 120

$$z = \frac{120 - 100}{10} = 2.00$$

Given score: 100

$$z = \frac{100 - 100}{10} = 0$$

Given score: 90

$$z = \frac{90 - 100}{10} = -1.00$$

Do Exercises 13 through 17 at the right.

Interpreting z-scores as Percentile Ranks

We are now in the position to use the normal curve to interpret these z-scores. To do so, we make use of the following table, which shows the percentage of area under

z	(A) Percentile rank (percent of area below z)	(C) Percent of area beyond z	z	(A) Percentile rank (percent of area below z)	(C) Percent of area beyond z
−2.2	1	99	0.1	54	46
−2.1	2	98	0.2	58	42
−2.0	2	98	0.3	62	38
−1.9	3	97	0.4	66	34
−1.8	4	96	0.5	69	31
−1.7	4	96	0.6	73	27
−1.6	5	95	0.7	76	24
−1.5	7	93	0.8	79	21
−1.4	8	92	0.9	82	18
−1.3	9	91	1.0	84	16
−1.2	12	88	1.1	86	14
−1.1	14	86	1.2	88	12
−1.0	16	84	1.3	91	9
−0.9	18	82	1.4	92	8
−0.8	21	79	1.5	93	7
−0.7	24	76	1.6	95	5
−0.6	27	73	1.7	96	4
−0.5	31	69	1.8	96	4
−0.4	34	66	1.9	97	3
−0.3	38	62	2.0	98	2
−0.2	42	58	2.1	98	2
−0.1	46	54	2.2	99	1
0.0	50	50			

Assume a mean of 50 and a standard deviation of 5; transform the following scores to z-scores.

13. 60

14. 45

15. 50

16. 65

17. 35

the normal curve. This table is abbreviated in order to simplify the present exposition. A complete presentation of percent of area above and below a given z appears in the Appendix in Table A-1.

Using this table, we can express any z-score as a percentile rank. To illustrate, in a prior example in which the mean was 100 and the standard deviation was 10, a score of 120 was found to have a corresponding z of 2.00. Looking at 2.0, in column (A) of the table below we find the percentile rank (percent of area *below z*) to be 98. In column (C) we find that two percent of the area is beyond a z of 2.0.

Examples

The percentile rank of a z of −1.00 is 16.

The percentile rank of a z of 0 is 50.

The percentile rank of a z of −0.6 is 27.

Do Exercises 18 through 27 at the right.

Let us now combine the two steps involved in interpreting a value of a normally distributed variable. If the mean is 500 and the standard deviation is 100, what is the percentile rank of a score of 560?

$$z = \frac{560 - 500}{100} = 0.6$$

Using the table above, we find the percentile rank of a score of 560 to be 73.

Examples

The percentile rank of a score of 480 is

$$z = \frac{480 - 500}{100} = -0.2.$$

The percentile rank of $z = -0.2$ is 42. Thus, the percentile rank of 480 is 42.

The percentile rank of a score of 360 is

$$z = \frac{360 - 500}{100} = -1.4.$$

The percentile rank of $z = -1.4$ is 8. Thus, the percentile rank of 360 is 8.

Do Exercises 28 through 35 at the right.

Use the accompanying table to find the percentile rank of the following z-scores.

18. z = −2.2

19. z = −1.5

20. z = −1.4

21. z = −0.8

22. z = −0.2

23. z = 0.1

24. z = 0.8

25. z = 1.7

26. z = 1.9

27. z = 2.2

A normally distributed variable has a mean of 100 and a standard deviation of 16. Find the percentile rank of the following scores. (Round z to one decimal place.)

28. 85

29. 125

30. 74

31. 108

32. 94

33. 118

34. 89

35. 135

Examples

Use Table A-1 in the Appendix to find the percentile rank of the following scores in which the mean is 70 and the standard deviation is 12.

If $X = 62$,

$$z = \frac{62 - 70}{12} = -0.67.$$

The percentile rank, found under column (A), is 25.14.

If $X = 79$,

$$z = \frac{79 - 70}{12} = 0.75.$$

The percentile rank of a score of 79 is 77.34.

Do Exercises 36 through 39 at the right.

Finding the Percent of the Area Between Scores

We are frequently called upon to find the percent of the area between two scores. In some cases, both scores are below the mean; in others, both are above the mean. In yet others, one is above the mean and the other is below the mean.

In the following illustrations, we shall use a mean of 100 and a standard deviation of 16.

Both scores below the mean:

A. Find the percentile rank of the score closer to the mean.
B. Find the percentile rank of the remaining score.
C. Subtract the second from the first.

Examples

Given: $X_1 = 85$, $X_2 = 95$.

$$z_{X_1} = \frac{85 - 100}{16} = -0.94 \qquad z_{X_2} = \frac{95 - 100}{16} = -0.31$$

Percentile rank of -0.94 is 17.36.
Percentile rank of -0.31 is 37.83.
Percent of area between scores of 85 and 95 is

$$37.83 - 17.36 = 20.47.$$

Use Table A-1 in the Appendix to find the percentile rank of the indicated scores. The mean is 84 and the standard deviation is 6.

36. 85

37. 92

38. 73

39. 68

Given: $X_1 = 70$, $X_2 = 99$.

$$z_{X_1} = \frac{70 - 100}{16} = -1.88 \qquad z_{X_2} = \frac{99 - 100}{16} = -0.06$$

Percentile rank of -1.88 is 3.01.
Percentile rank of -0.06 is 47.61.
Percent of area between scores of 70 and 99 is

$47.61 - 3.01 = 44.60$.

Do Exercises 40 through 43 at the right.

Both scores above the mean:

A. Find the percentile rank of the higher score—the one farther from the mean.
B. Find the percentile rank of the lower score—the one closer to the mean.
C. Subtract the second from the first.

Example

Given: $X_1 = 128$, $X_2 = 106$.

$$z_{X_1} = \frac{128 - 100}{16} = \frac{28}{16} = 1.75 \quad z_{X_2} = \frac{106 - 100}{16} = \frac{6}{16} = 0.38$$

Percentile rank of 1.75 is: 95.99
Percentile rank of 0.38 is: 64.80
Subtracting gives: 31.19

The percent of area between the two scores is 31.19.

Do Exercises 44 through 47 at the right.

One score above the mean, the other below:

A. Find the percentile rank of the score above the mean.
B. Find the percentile rank of the score below the mean.
C. Subtract the second from the first.

Example

Given: $X_1 = 131$, $X_2 = 92$.

$$z_{X_1} = \frac{131 - 100}{16} = 1.94 \qquad z_{X_2} = \frac{92 - 100}{16} = -0.50$$

Percentile rank of 1.94 is: 97.38
Percentile rank of -0.50 is: 30.85
Subtracting gives: 66.53

The percent of the area between the two scores is 66.53.

Do Exercises 48 through 50 at the right.

Find the percent of the area between the indicated scores. The mean is 100 and the standard deviation is 16.

40. $X_1 = 89$, $X_2 = 94$

41. $X_1 = 77$, $X_2 = 79$

42. $X_1 = 96$, $X_2 = 98$

43. $X_1 = 99$, $X_2 = 65$

Find the percent of the area between the indicated scores. The mean is 100 and the standard deviation is 16.

44. $X_1 = 136$, $X_2 = 116$

45. $X_1 = 105$, $X_2 = 101$

46. $X_1 = 137$, $X_2 = 104$

47. $X_1 = 117$, $X_2 = 110$

Find the percent of the area between the indicated scores. The mean is 100 and the standard deviation is 16.

48. $X_1 = 124$, $X_2 = 93$

49. $X_1 = 103$, $X_2 = 97$

50. $X_1 = 117$, $X_2 = 84$

Generalizing: To find the percent of the area between two z-scores, find the percentile rank of the higher score and subtract from it the percentile rank of the lower score. [Note: $z = -1.00$ is higher than $z = -1.50$ which, in turn, is higher than $z = -2.50$, etc.]

Examples

Find the percent of the area between $z = -1.83$ and $z = 1.43$. Since $z = 1.43$ is higher than $z = -1.83$, find percentile rank corresponding to $z = 1.43$ and subtract from it percentile rank corresponding to $z = -1.83$.

Percentile rank of $z = 1.43$ is:	92.36
Percentile rank of $z = -1.83$ is:	3.36
Percent of area between is:	89.00

Find the percent of the area between $z = -1.83$ and $z = -2.05$.

Percentile rank of $z = -1.83$ is:	03.36
Percentile rank of $z = -2.05$ is:	02.02
Percent of area between is:	1.34

COMPARING SCORES ON MORE THAN ONE VARIABLE

It has often been said that you cannot compare apples and applecarts. This statement is not true if both variables are normally distributed and you know the mean and standard deviation of each.

The z-scores represent an individual's (or object's) relative position on a given variable. The higher the z, the higher that person's relative position and the higher his or her percentile rank.

To illustrate, Mary B. obtained a score of 130 on variable X that has a mean of 100 and a standard deviation of 16. She obtained a score of 620 on variable Y that has a mean of 500 and a standard deviation of 100. On which of these variables did she score higher?

Her z and corresponding percentile rank on X is

$$z_X = \frac{130 - 100}{16} = 1.88.$$

Percentile rank equals 96.99.

Her performance on variable Y is:

$$z_Y = \frac{620 - 500}{100} = 1.20$$

Percentile rank equals 88.49.

7

Since her percentile rank on variable X is greater than her percentile rank on variable Y, she scored higher on X.

Example

Thomas A. obtained a score of 120 on X and 704 on Y. On which variable did he score higher?

$$z_X = \frac{120 - 100}{16} = 1.25$$

Percentile rank equals 89.44.

$$z_Y = \frac{704 - 500}{100} = 2.04$$

Percentile rank equals 97.93.

Thomas A. scored higher on variable Y.

Do Exercises 51 through 55 at the right.

The Coefficient of Variation

The standard deviation may also be thought of as an estimate of precision in measurement. The coefficient of variation, V, employs the standard deviation to express variation relative to the magnitude of the values of a variable. To illustrate, a market analyst may be interested in learning if a high-priced stock shows greater or lesser variation in price, relative to its magnitude, than a low-priced stock.

The coefficient of variation is defined as follows:

$$V = \frac{100s}{\bar{X}}$$

Example

Over the past years, stock A has had a mean selling price of \$435 with a standard deviation (based upon price at the end of each day of trading) of \$10. Stock B has had a mean selling price of \$2.35, with a standard deviation of 0.15. Which stock shows the greater relative variability?

$$V_A = \frac{10 \times 100}{435} = \frac{1000}{435} = 2.30$$

$$V_B = \frac{0.15 \times 100}{2.35} = \frac{15}{2.35} = 6.38$$

Conclusion: The low priced stock showed the greater relative variation over the course of a year of trading.

Do Exercises 56 through 60 at the right.

Variable X has an associated mean of 50 and a standard deviation of 10. The mean and standard deviation on variable Y are 15 and 3, respectively. Determine which of the paired scores obtained a higher percentile rank.

51. $X = 45$, $Y = 12$

52. $X = 63$, $Y = 20$

53. $X = 30$, $Y = 9$

54. $X = 52$, $Y = 17$

55. $X = 48$, $Y = 13$

Find the relative variation of the following variables.

56. $\bar{X}_A = 65$, $s_A = 6$

57. $\bar{X}_B = 440$, $s_B = 12$

58. $\bar{X}_C = 800$, $s_C = 9$

59. $\bar{X}_D = 15$, $s_D = 0.20$

60. $\bar{X}_E = 6$, $s_E = 0.15$

CHAPTER 7 TEST

Use the data given in each problem and Table A-1 in the Appendix to complete this test.

1. The mean equals 6.09, the standard deviation equals 1.22, and the variance is 1.49. The z corresponding to $X = 5.62$ is:

   a) 4.61   b) −0.38   c) −4.61   d) 0.38

2. The mean is 50, the standard deviation is 3, and the variance is 9. The z corresponding to $X = 54$ is:

   a) 1.33   b) 8.00   c) 0.44   d) −0.44

3. The mean is 100, the standard deviation is 14, and the variance is 196. The z corresponding to $X = 82$ is:

   a) 5.86   b) 1.29   c) 0.42   d) −1.29

4. The mean is 0.76, the standard deviation is 0.08 and the variance is 0.0064. The z corresponding to $X = 0.82$ is:

   a) 7.5   b) 0.75   c) 0.08   d) 1.08

5. The mean is 250, the standard deviation is 21, and the variance is 441. The z corresponding to $X = 219$ is:

   a) −0.88   b) 1.48   c) 0.88   d) −1.48

6. The mean is 16, the standard deviation is 3, and the variance is 9. The z corresponding to $X = 20$ is:

   a) 1.33   b) −1.33   c) −5.33   d) −0.44

7. The mean is 84, the standard deviation is 6, and the variance is 36. The z corresponding to $X = 100$ is:

   a) −2.67   b) 1.25   c) −1.25   d) 2.67

8. The mean is 400, the standard deviation is 40, and the variance is 1600. The z corresponding to $X = 294$ is:

   a) 1.36   b) −2.65   c) −1.36   d) 2.65

9. The mean is 55, the standard deviation is 7, and the variance is 49. The z corresponding to $X = 56$ is:

   a) 0.98   b) 7.86   c) 1.02   d) 0.14

10. Given $z = -1.63$, the corresponding percentile rank is:

    a) 94.84   b) 44.84   c) 5.16   d) 0.1074

11. Given $z = 0.42$, the corresponding percentile rank is:

    a) 66.28   b) 33.72   c) 16.28   d) 51.60

12. Given $z = -0.09$, the corresponding percentile rank is:

    a) 03.59   b) 46.41   c) 53.59   d) 18.41

13. The following scores were made by Person C on four scales.

Scale	X	Mean	Standard deviation
1	150	100	25
2	120	100	7
3	490	500	100
4	49	50	4

Relative to the mean and standard deviation, the lowest score was made on:

a) Scale 1   b) Scale 2   c) Scale 3   d) Scale 4

14. The highest score in the above example was made on:

a) Scale 1   b) Scale 2   c) Scale 3   d) Scale 4

15. On a scale in which the mean is 100 and the standard deviation is 12, the percent of area between scores of 92 and 105 is:

a) 41.14   b) 85.99   c) 35.99   d) 91.42

16. On a scale in which the mean is 50 and the standard deviation is 5, the percent of area between scores of 54 and 62 is:

a) 94.52   b) 44.52   c) 20.37   d) 19.63

17. On a scale in which the mean is 500 and the standard deviation is 60, the area between scores of 390 and 380 is:

a) 53.98   b) 03.98   c) 0.08   d) 5.64

18. The area between $z = 1.87$ and $z = -1.02$ is:

a) 30.23   b) 81.54   c) 80.23   d) 99.81

19. The area between $z = -0.87$ and $z = -1.53$ is:

a) 00.82   b) 24.86   c) 12.92   d) 74.54

20. Following a survey of forest resources in four geographical regions in the United States, the mean height and standard deviation of native trees were determined. The trees in which region evidenced the greater relative variability?

Region	Mean height	Standard deviation
A	46	12
B	88	15
C	30	9
D	15	4

a) Region A   b) Region B   c) Region C   d) Region D

Practical

21. Find the z and percentile ranks of the following scores of normally distributed variables.

	X	Mean	Standard deviation
a)	18	24	9
b)	29	24	9
c)	36	24	9
d)	35	50	6
e)	42	50	6
f)	63	50	6
g)	102	100	15
h)	97	100	15
i)	78	100	15

22. Following are scores made by two individuals, E and F, on a number of different scales. Indicate who made the higher scores on each scale.

	E			F		
	X	Mean	Standard deviation	X	Mean	Standard deviation
a)	60	80	8	80	100	10
b)	150	100	16	720	500	100
c)	75	50	15	130	100	12
d)	90	100	8	380	500	100

23. Find the coefficient of variation (V) for each of the following

a) $\bar{X} = 7.83$, $s = 1.04$

b) $\bar{X} = 0.94$, $s = 0.16$

c) $\bar{X} = 23.52$, $s = 4.64$

d) $\bar{X} = 88.76$, $s = 12.45$

Transforming Nonnormal Distributions of Scores into Normally Distributed Variables

In Chapter 7, we learned how to interpret scores when we know the mean and standard deviation of a normally distributed variable. Most empirical distributions are not normally distributed, even if relatively bell-shaped. (Standardized psychological and educational tests are normally distributed because scores have been transformed into units of the standard normal curve prior to the publication of test norms). There is usually some degree of skewness. The standard normal curve cannot be used to interpret, with a great degree of precision, scores of a variable that is not normally distributed.

In this chapter we will learn the procedures for transforming nonnormal distributions into normally distributed scores. The transformed scores have a mean of 0 and a standard deviation of 1. The standard normal curve may then be used to interpret these transformed scores. We will also learn the T-score transformation which permits us to express each transformed score as a positive integer.

TRANSFORMING NONNORMAL FREQUENCY DISTRIBUTIONS INTO AREAS OF THE STANDARD NORMAL CURVE

There are several different techniques for normalizing a frequency distribution of scores. All yield more or less the same results. The method shown here is the one preferred by the author.

To illustrate the procedures for normalizing frequency distributions, we shall employ the grouped frequency distribution ($i = 2$) of sixty miles-per-gallon ratings of 1976 cars. These are reproduced below. Note the extreme positive skew.

Real limits of class interval	f	Cum f
27.5–29.5	1	60
25.5–27.5	1	59
23.5–25.5	2	58
21.5–23.5	3	56
19.5–21.5	7	53
17.5–19.5	6	46
15.5–17.5	10	40
13.5–15.5	6	30
11.5–13.5	10	24
9.5–11.5	14	14

OBJECTIVES

- Know how to transform nonnormal frequency distributions into normally distributed variables with a mean of 0 and a standard deviation of 1.

- Know how to prepare a graph of a normalized frequency distribution of scores.

- Know how to convert these transformed scores into T-scores.

8

Step A. List all the integers from the highest in the upper class interval ($X = 29$) to the lowest in the bottom class interval ($X = 10$). See column (A) of the Table 8-1 below.

Step B. Add 1 to the score near the upper real limit of the highest interval (29.4) and to the lower real limit of the lowest interval (9.6). This is done to provide "anchor" points for drawing the graph.

Step C. Find the score with a corresponding percentile rank of 50. In the present example:

a) $50 \times 60 \div 100 = 30$

b) 30th score is found at the upper real limit of the interval 13.5–15.5. Thus, the score at the 50th percentile is 15.5. This score will correspond to the mean, median, and mode in the normalized distribution. It should be placed between 15 and 16 in the following table.

	Score	Percentile rank	Corresponding z under standard normal curve	Height of ordinate
	29.4			
	29			
	28			
	27			
	26			
	25			
	24			
	23			
	22			
	21			
	20			
	19			
	18			
	17			
Mode	16			
Mean	15.5			
Median	15			
	14			
	13			
	12			
	11			
	10			
	9.6			

Do Exercise 1 at the right.

EXERCISES

1. Shown below is the grouped frequency distribution of sixty-three scores. Follow steps A through C.

Real limits of class interval ($i = 3$)	f	Cum f
24.5–27.5	1	63
21.5–24.5	3	62
18.5–21.5	8	59
15.5–18.5	11	51
12.5–15.5	16	40
9.5–12.5	10	24
6.5– 9.5	9	14
3.5– 6.5	4	5
0.5– 3.5	1	1
	63	

Step D. Use the procedure learned in Chapter 4 to find the percentile rank of each of the scores.

Examples

Percentile rank of score of 29.4 is

$$\frac{\left(\frac{1.9}{2}\right)1 + 59}{60} \times 100 = \frac{59.95}{60} \times 100$$

$$= 99.92.$$

Percentile rank of score of 19 is

$$\frac{\left(\frac{1.5}{2}\right)6 + 40}{60} \times 100 = \frac{44.5}{60} \times 100$$

$$= 74.17$$

Table 8-1

(A) Score	(B) Percentile rank	(C) Corresponding z under standard normal curve	(D) Height of ordinate
29.4	99.92	3.14	0.0029
29	99.58	2.64	0.0122
28	98.75	2.24	0.0325
27	97.92	2.04	0.0498
26	97.08	1.89	0.0669
25	95.83	1.73	0.0893
24	94.17	1.57	0.1163
23	92.08	1.41	0.1476
22	89.58	1.26	0.1804
21	85.42	1.05	0.2299
20	79.58	0.83	0.2827
19	74.17	0.65	0.3230
18	69.17	0.50	0.3521
17	62.50	0.32	0.3790
16	54.17	0.10	0.3970
15.5	50.00	0.00	0.3989
15	47.50	−0.06	0.3988
14	42.50	−0.19	0.3918
13	35.83	−0.36	0.3739
12	27.50	−0.60	0.3332
11	17.50	−0.93	0.2589
10	05.83	−1.57	0.1163
9.6	01.17	−2.27	0.0303

Do Exercises 2 through 31 at the right.

Find the percentile rank of the following values.

	(A) Score	(B) Percentile rank
2.	27.4	
3.	27	
4.	26	
5.	25	
6.	24	
7.	23	
8.	22	
9.	21	
10.	20	
11.	19	
12.	18	
13.	17	
14.	16	
15.	15	
16.	14	
17.	13.81	
18.	13	
19.	12	
20.	11	
21.	10	
22.	9	
23.	8	
24.	7	
25.	6	
26.	5	
27.	4	
28.	3	
29.	2	
30.	1	
31.	0.6	

8

Step E. Turn to Table A-1. Find in the body of the table the percentile rank that most closely approximates the percentile ranks listed in column (B) of Table 8-1. For example, the score of 12 has a percentile rank of 27.50. The closest value to this is 27.43. The z corresponding to 27.43 is -0.60. Record each z in column (C).

Examples

1. A score of 28 is at the 98.75th percentile. The z corresponding to this percentile rank is 2.24.
2. A score of 15 is at the 47.50th percentile. The value closest to this in the body of the table is 47.61. This corresponds to a z of -0.06.

Do Exercises 32 through 61 at the right.

TRANSFORMING TO *T*-SCORES

We have now transformed the nonnormal frequency distribution into a normal distribution with a mean of 0 and a standard deviation of 1. However, most scores involve decimal values and all scores below the mean are negative. It is usually desirable to express values for a normally distributed variable in terms of a positive number. This can be accomplished by a simple transformation:

$$T = a + bz$$

The selection of the constants a and b depends upon which mean and standard deviation you want in the final transformed distribution. For example, if you want the mean of the transformed distribution to be 500, you set a equal to 500. If you want the scores of the transformed distribution to be a whole number, you set b equal to 100 or more.

Examples

Show the *T*-transformation necessary to produce a mean equal to 50 and a standard deviation equal to 5.

$$T = a + bz$$
$$= 50 + 5z$$

Show the transformation necessary to produce a mean of 500 and a standard deviation of 100.

$$T = a + bz$$
$$= 500 + 100z$$

Find the z-scores most closely approximating each percentile rank and record in column (C).

	(A) Score	(B) Percentile rank	(C) Corresponding z
32.	27.4	99.95	
33.	27	99.70	
34.	26	99.20	
35.	25	98.68	
36.	24	97.62	
37.	23	96.03	
38.	22	94.44	
39.	21	91.54	
40.	20	87.30	
41.	19	83.06	
42.	18	78.05	
43.	17	72.22	
44.	16	66.40	
45.	15	59.25	
46.	14	50.79	
47.	13.81	50.00	
48.	13	42.33	
49.	12	35.44	
50.	11	30.16	
51.	10	24.87	
52.	9	19.84	
53.	8	15.08	
54.	7	10.32	
55.	6	06.87	
56.	5	04.76	
57.	4	02.65	
58.	3	01.32	
59.	2	00.79	
60.	1	00.27	
61.	0.6	00.05	

Do Exercises 62 through 66 at the right.

Let us now transform the miles-per-gallon scores into a T-score so that we obtain a mean equal to 50 and a standard deviation equal to 10.

A. Take the z corresponding to each score and multiply by $b = 10$.

Examples

The z of a score of 29.4 is 3.14.

$b \times 3.14 = 10 \times 3.14$

$\qquad = 31.4$

The z of the mean score, 15.5, is 0.

$b \times 0 = 10 \times 0$

$\qquad = 0$

B. Add the constant $a = 50$ to each bz to yield the transformed score, T.

The score of 29.4 becomes

$T = a + bz$

$\qquad = 50 + 31.4$

$\qquad = 81.4.$

(Note: In many practical applications of the T-score transformation, the number is rounded to the nearest integer so that the T is expressed as a positive *whole* number.)

Examples

The T of the mean becomes

$T = a + bz$

$\quad = 50 + 0$

$\quad = 50.$

The T of a score of 11 becomes

$T = a + bz$

$\quad = 50 + 10 \, (-0.93)$

$\quad = 40.7.$

Show T-transformations necessary to yield the following means and standard deviations:

62. $\bar{X} = 70$, $s = 15$

63. $\bar{X} = 50$, $s = 5$

64. $\bar{X} = 100$, $s = 16$

65. $\bar{X} = 500$, $s = 50$

66. $\bar{X} = 10$, $s = 2$

Shown below are the *T*-score transformations of all the normalized values of the miles-per-gallon data.

Score	Corresponding z under standard normal curve	*T*-transformation ($T = 50 + 10z$)	Rounded to nearest whole number
29.4	3.14	81.4	81
29	2.64	76.4	76
28	2.24	72.4	72
27	2.04	70.4	70
26	1.89	68.9	69
25	1.73	67.3	67
24	1.57	65.7	66
23	1.41	64.1	64
22	1.26	62.6	63
21	1.05	60.5	60
20	0.83	58.3	58
19	0.65	56.5	56
18	0.50	55.0	55
17	0.32	53.2	53
16	0.10	51.0	51
15.5	0.00	50.0	50
15	−0.06	49.7	50
14	−0.10	48.1	48
13	−0.36	46.4	46
12	−0.60	44.0	44
11	−0.93	40.7	41
10	−1.57	34.3	34
9.6	−2.27	27.3	27

Do Exercises 67 through 96 at the right.

Any transformed score can be readily interpreted by converting to units of the standard normal curve.

Examples

1. A score of 81.9 yields the following:

$$z = \frac{81.9 - 50}{10} = \frac{31.9}{10}$$

$$= 3.19$$

Table A-1 reveals that the corresponding percentile rank is 99.93.

2. A score of 37.8 yields the following:

$$z = \frac{37.8 - 50}{10} = \frac{-12.2}{10}$$

$$= -1.22$$

Table A-1 reveals that the corresponding percentile rank is 11.12.

Convert the following scores to *T*-scores, using the transformation $T = 500 + 100z$.

	Score	Corresponding z under standard normal curve	*T*-transformation ($T = 500 + 100z$)
67.	27.4	3.30	
68.	27	2.75	
69.	26	2.41	
70.	25	2.22	
71.	24	1.98	
72.	23	1.75	
73.	22	1.59	
74.	21	1.37	
75.	20	1.14	
76.	19	0.96	
77.	18	0.77	
78.	17	0.59	
79.	16	0.42	
80.	15	0.23	
81.	14	0.02	
82.	13.81	0.00	
83.	13	−0.19	
84.	12	−0.37	
85.	11	−0.52	
86.	10	−0.68	
87.	9	−0.85	
88.	8	−1.03	
89.	7	−1.26	
90.	6	−1.49	
91.	5	−1.67	
92.	4	−1.94	
93.	3	−2.22	
94.	2	−2.42	
95.	1	−2.78	
96.	0.6	−3.30	

Do Exercises 97 through 101 at the right.

PLOTTING A GRAPH OF A NORMALIZED DISTRIBUTION

To illustrate the procedures for constructing a graph of a normalized distribution of z-scores, we shall use the previously calculated transformed miles-per-gallon scores. These are reproduced below.

(A) Score	(B) Corresponding z under standard normal curve	(C) Height of ordinate
29.4	3.14	
29	2.64	
28	2.24	
27	2.04	
26	1.89	
25	1.73	
24	1.57	
23	1.41	
22	1.26	
21	1.05	
20	0.83	
19	0.65	
18	0.50	
17	0.32	
16	0.10	
15.5	0.00	
15	−0.06	
14	−0.19	
13	−0.36	
12	−0.60	
11	−0.93	
10	−1.57	
9.6	−2.27	

A. Refer to column (D) of Table A-1. This shows the height of ordinate corresponding to each z-score. Find the height of the ordinate corresponding to each z in the preceding table and place in column (C).

Use data in Exercises 67 through 96 to find the percentile rank of the following scores.

97. 300

98. 658

99. 260

100. 790

101. 480

8

Examples

1. The height of the ordinate corresponding to $z = 3.14$ is 0.0029.
2. The height of the ordinate corresponding to $z = 0$ is 0.3989.
3. The height of the ordinate corresponding to $z = -1.57$ is 0.1163.

Do Exercises 102 through 131 at the right.

B. On a piece of graph paper, draw the horizontal or X-axis so that values between -3.4 through 3.4 are equally spaced.

C. Draw the vertical or Y-axis so that two-placed decimal values from 0.00 to 0.40 are represented.

D. Locate each z score along the horizontal axis and move vertically until you find the corresponding value representing the height of the ordinate. Place a dot at this point.

E. When you have completed all the values and joined the dots, you will have a normal distribution of your transformed scores. You may also add the corresponding original scores and their transformed T-scores to the legend along the X-axis. (See figure below).

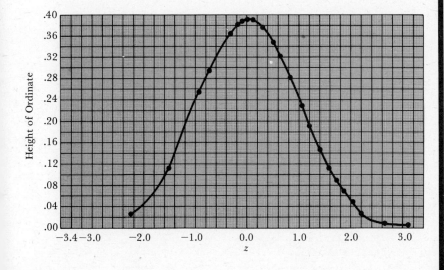

Do Exercise 132 at the right. Use the grid on p. 104.

Find and record the height of the ordinate corresponding to the following z-scores.

	(A) Score	(B) Corresponding z	(C) Height of ordinate
102.	27.4	3.30	
103.	27	2.75	
104.	26	2.41	
105.	25	2.22	
106.	24	1.98	
107.	23	1.75	
108.	22	1.59	
109.	21	1.37	
110.	20	1.14	
111.	19	0.96	
112.	18	0.77	
113.	17	0.59	
114.	16	0.42	
115.	15	0.23	
116.	14	0.02	
117.	13.81	0.00	
118.	13	−0.19	
119.	12	−0.37	
120.	11	−0.52	
121.	10	−0.68	
122.	9	−0.85	
123.	8	−1.03	
124.	7	−1.26	
125.	6	−1.49	
126.	5	−1.67	
127.	4	−1.94	
128.	3	−2.22	
129.	2	−2.42	
130.	1	−2.78	
131.	0.6	−3.30	

132. Prepare a graph of the transformed scores of Exercises 102 through 131.

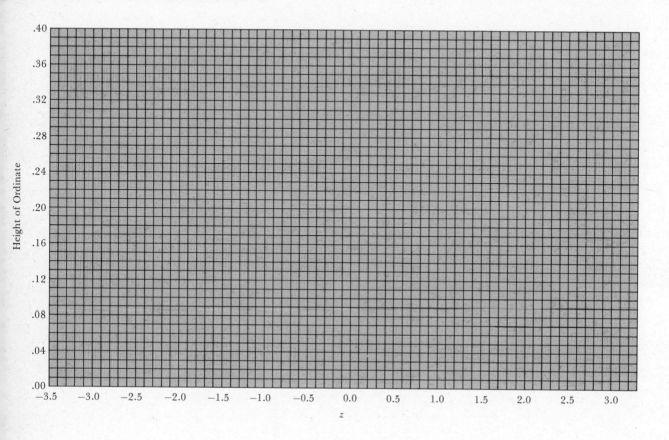

CHAPTER 8 TEST

Use the data given in each problem and Table A-1 in the Appendix to complete this test.

1. Given the percentile rank of a score equals 03.14, the corresponding z is:

    a) 00.08   b) 49.92   c) −1.86   d) 96.86

2. Given the percentile rank of a score equals 67.00, the corresponding z is:

    a) 0.44   b) 25.14   c) 74.86   d) 0.3187

3. Given the percentile rank of a score equals 24.20, the corresponding z is:

    a) 0.70   b) 0.65   c) −1.00   d) −0.70

4. Given the z corresponding to a score equals −2.62, the height of the ordinate is:

    a) −1.94   b) 0.0608   c) 0.0129   d) 00.44

5. Given the z corresponding to a score equals 1.79, the height of the ordinate is:

    a) 96.33   b) 0.0804   c) −2.10   d) 0.0440

6. Given the z corresponding to a score equals 0.01, the height of the ordinate is:

    a) 0.3989   b) 0.0004   c) 49.99   d) 0.0035

7. If we wanted to set the standard deviation equal to 10 in the transformation $T = a + bz$, we would:

    a) assign a value of 10 to all values of z

    b) assign a value of 10 to b

    c) assign a value of 10 to a

    d) assign a value of 10 to both a and b

8. If we wanted to set the mean equal to 40 in the transformation $T = a + bz$, we would:

    a) assign a value of 40 to a

    b) assign a value of 40 to z

    c) assign a value of 40 to b

    d) add 40 to all values of z and divide by N

9. If a transformation $T = 280 + 15z$ is used:

    a) the mean is equal to 15 and the standard deviation is equal to 280

    b) the mean is equal to 280 but the standard deviation cannot be known without knowing z

c) the mean is equal to 280 and the standard deviation is equal to $15z$

d) the mean is equal to 280 and the standard deviation is equal to $15z$

10. The transformation necessary to yield a mean equal to 430 and a standard deviation equal to 75 is:

a) $T = 75 + 430z$ b) $T = 430 + 75z$

c) $T = 430 + 75b$ d) $T = 75 + 430b$

11. Using the transformation $T = 160 + 14z$, a standard score of 0.17 would have a corresponding T value of:

a) 183.80   b) 2.38   c) 174.00   d) 162.38

12. Using the transformation $T = 120 + 16z$, a standard score of -0.11 would have a corresponding z of:

a) 122.76   b) 98.24   c) 118.24   d) 102.40

13. Using the transformation $T = 170 + 8z$, a z of -0.76 would have a corresponding value of:

a) 163.92   b) 162.00   c) 176.08   d) 109.20

14. If a given T equals 740 in a distribution with a standard deviation equal to 100 and a mean equal to 500, the corresponding z is:

a) 0.24   b) 7.40   c) 2.40   d) -0.24

15. If a given T equals 46 in a distribution with a standard deviation equal to 4 and a mean equal to 55, the corresponding percentile rank is:

a) 2.25   b) 01.22   c) 98.78   d) 15.87

Practical

16. Transform the scores in the following grouped frequency distribution into units of the standard normal curve. (To reduce calculations, select all odd numbered integers from 9 through 47.)

Class interval	f	Cum f
44–47	1	40
40–43	2	39
36–39	2	37
32–35	4	35
28–31	4	31
24–27	5	27
20–23	8	22
16–19	5	14
12–15	5	9
8–11	4	4
	$N = 40$	

17. Convert the z-scores to T's, using the transformation $T = 300 + 100z$.

18. Prepare a graph of the T-distribution.

Correlation for Quantitative and Ordinal Variables

Up to this point in the course, we have been looking at the various descriptive measures available for use with single variables—e.g., measures of central tendency and dispersion. In this chapter we investigate several measures that are available to express, quantitatively, the relationship between variables. Measures that describe how two variables vary together are known as *correlation coefficients.*

Although there are a number of different types of correlation coefficients, all have the following characteristics in common: (a) Two sets of measurements are obtained on the same individual, objects, or events. (b) The values of the correlation coefficients vary between +1.00 and −1.00. Both of these extremes represent perfect relationships between the variables, and a coefficient of 0.00 represents the absence of a relationship. (c) A positive relationship means that individuals, objects, or events obtaining high scores on one variable tend to obtain high scores on the second variable and, conversely, those scoring low on one variable tend to score low on a second variable. (d) A negative relationship means that high scores on one variable tend to be accompanied by low scores on a second and, conversely, low scores on one variable tend to accompany high scores on a second variable.

CONSTRUCTING A SCATTER DIAGRAM

Visual inspection of a scatter diagram will usually permit you to judge if two variables are related, how they are related (positively or negatively), and if the relationship is weak (low correlation) or strong (high correlation).

The following table shows both cubic inches of engine displacement and highway miles per gallon of twenty 1976 automobiles with manual transmission. Note that there are two "scores" for each car. We shall call the cubic inches of engine displacement the *X*-variable and the miles-per-gallon scores the *Y*-variable.

OBJECTIVES

- Know how to construct a scatter diagram for presenting visually the relationship between two variables.

- Know how to calculate Pearson *r* for quantitative variables.

- Know the various steps for calculating Spearman rho (r_{rho}) for ordinally scaled data.

  a) Know how to convert quantitative values of a variable to ranks.

  b) Know how to calculate r_{rho} when both variables are expressed as ranks.

  c) Know how to apply the Pearson *r* formula to ranked data when there are tied ranks.

Make of car	(X) Cubic inches of engine displacement	(Y) Highway miles per gallon
Gremlin	236	26
Hornet	304	16
Audi Fox	97	36
Skyhawk	231	26
Vega	140	30
Datsun B-210	85	41
Datsun 280Z	168	25
Fiat 131	107	30
Mustang II	171	24
Granada	250	20
Capri II	171	27
Monarch	250	20
Starfire	231	26
Val/Duster	225	23
Sunbird	231	26
Lemans	260	21
Corolla	97	35
Corona Mark II	156	22
Volvo	240	27
LUV-pickup	111	32

Step-by-Step Procedures

A. On a piece of graph paper, mark equally spaced intervals on the X-axis going from 85 (the lowest cu. in. displacement) to 305 (approximately the highest cu. in. displacement in the sample).

B. Do the same with the Y-axis, showing equally spaced intervals between 16 mpg and 41 mpg.

C. Look at the first car, Gremlin. Locate the value 236 on the Y-axis. Move vertically from this point until it intercepts the line representing 26 mpg on the Y-axis. Place a dot at this point.

D. Repeat the procedures in step C until you have plotted the paired scores of all twenty automobiles.

The following figure shows the plot of all twenty paired scores. By visual inspection it can be seen that the dots form a generally descending pattern from the upper left-hand corner to the lower right-hand corner. The correlation is, therefore, negative. Subsequent calculation of the coefficient will show the correlation to be high.

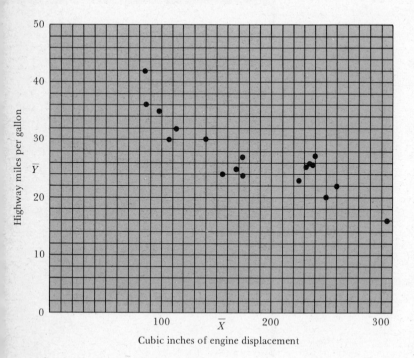

Cubic inches of engine displacement

Do Exercise 1 at the right.

CORRELATION COEFFICIENT WITH QUANTITATIVE VARIABLES—Pearson *r*

Pearson *r* is defined as

$$\frac{\Sigma xy}{\sqrt{\Sigma x^2 \cdot \Sigma y^2}}.$$

The only new term in the formula is Σxy—the sum of the cross products.

Σx^2* refers to the sum of squares of the *X*-variable and Σy^2 is the sum of squares of the *Y*-variable. Let us review the calculation of these sums of squares with the following set of data.

---

* In Chapter 6, the symbol for sum of squares was given as $\Sigma(X - \bar{X})^2$. Most authors prefer the shortened notation Σx^2. From this point forward, we shall use Σx^2 to designate the sum of squares of variable *X* and Σy^2 to designate the sum of squares of variable *Y*.

X	Y	X	Y
236	19	171	23
304	20	250	21
97	32	231	24
231	24	225	20
140	27	231	24
85	35	260	18
168	22	97	26
107	25	156	18
171	20	240	23
250	21	111	28

X	X²	Y	Y²
2	4	19	361
3	9	21	441
3	9	21	441
4	16	23	529
4	16	23	529
4	16	23	529
5	25	25	625
5	25	25	625
6	36	27	729
$\Sigma X = 36$	$\Sigma X^2 = 156$	$\Sigma Y = 207$	$\Sigma Y^2 = 4809$

$N = 9$

A. The sum of squares of X is

$$\Sigma x^2 = \Sigma X^2 - \frac{(\Sigma X)^2}{N}.$$

We know

$\Sigma X^2 = 156;$

and

$$\frac{(\Sigma X)^2}{N} = \frac{(36)^2}{9} = \frac{1296}{9}$$

$$= 144.$$

B. Substituting in the formula for the sum of squares of X gives

$$\Sigma x^2 = 156 - 144$$

$$= 12.$$

Do Exercise 2 at the right.

C. The sum of squares of Y is

$$\Sigma y^2 = \Sigma Y^2 - \frac{(\Sigma Y)^2}{N}.$$

We know

$\Sigma Y^2 = 4809;$

and

$$\frac{(\Sigma Y^2)}{N} = \frac{(207)^2}{9} = \frac{42,849}{9}$$

$$= 4761.$$

$$\Sigma y^2 = 4809 - 4761$$

$$= 48$$

2. Find the sum of squares of X for the following distribution of paired scores.

X	X²	Y	Y²
4		25	
5		23	
5		23	
6		21	
6		21	
6		21	
7		19	
7		19	
8		17	
$\Sigma X =$	$\Sigma X^2 =$	$\Sigma Y =$	$\Sigma Y^2 =$

$\Sigma X^2 =$

$\dfrac{(\Sigma X)^2}{N} =$

$\Sigma x^2 = \Sigma X^2 - \dfrac{(\Sigma X)^2}{N}$

$=$

$=$

Do Exercise 3 at the right.

D. The sum of the cross products is defined as

$$\Sigma xy = \Sigma XY - \frac{(\Sigma X)(\Sigma Y)}{N}.$$

To obtain the sum of the cross products, proceed as follows:

(1) X	(2) X^2	(3) Y	(4) Y^2	(5) XY
2	4	19	361	38
3	9	21	441	63
3	9	21	441	63
4	16	23	529	92
4	16	23	529	92
4	16	23	529	92
5	25	25	625	125
5	25	25	625	125
6	36	27	729	162
$\Sigma X = 36$	$\Sigma X^2 = 156$	$\Sigma Y = 207$	$\Sigma Y^2 = 4809$	$\Sigma XY = 852$

A. Add a column headed by XY to the above table.
B. Multiply each X (column 1) by its corresponding Y (column 3) and place in column (5).
C. Sum column (5) to obtain ΣXY. In the present example $\Sigma XY = 852$.
D. Multiply ΣX times ΣY and divide by N, the number of pairs. In the present problem

$$\frac{(\Sigma X)(\Sigma Y)}{N} = \frac{(36)(207)}{9}$$

$$= 828.$$

E. Substitute the values obtained in steps C and D in the formula for the cross products.

In the present example

$$\Sigma xy = \Sigma XY - \frac{(\Sigma X)(\Sigma Y)}{N}$$

$$= 852 - 828$$

$$= 24.$$

In the event of a positive correlation, the sum of the cross products is positive. The sum of the cross products is negative when the correlation is negative.

3. Find the sum of squares of Y for the following distribution of paired scores.

X	X^2	Y	Y^2
4	16	25	
5	25	23	
5	25	23	
6	36	21	
6	36	21	
6	36	21	
7	49	19	
7	49	19	
8	64	17	
$\Sigma X = 54$	$\Sigma X^2 = 336$	$\Sigma Y =$	$\Sigma Y^2 =$

$$\Sigma Y^2 =$$

$$\frac{(\Sigma Y)^2}{N} =$$

$$\Sigma y^2 = \Sigma Y^2 - \frac{(\Sigma Y)^2}{N}$$

$$=$$

$$=$$

Do Exercise 4 at the right.

Substitute Σxy, Σx^2, and Σy^2 in the formula for r and solve:

$$r = \frac{\Sigma xy}{\sqrt{\Sigma x^2 \cdot \Sigma y^2}}$$

$$= \frac{24}{\sqrt{12 \cdot 48}} = \frac{24}{\sqrt{576}}$$

$$= \frac{24}{24}$$

$$= 1.00$$

The correlation of 1.00 is perfect positive. It is the highest positive correlation possible.

A scatter diagram would show that all data points fit on a straight line going from the lower left hand to the upper right hand corner. Only when the correlation is perfect (positive or negative) will the data points form a straight line.

Example

Find the correlation between cubic inches of engine displacement and highway mileage for twenty 1976 model cars equipped with manual transmission.

X	X^2	Y	Y^2	XY
236	55696	26	676	6136
304	92416	16	256	4864
97	9409	36	1296	3492
231	53361	26	676	6006
140	19600	30	900	4200
85	7225	41	1681	3485
168	28224	25	625	4200
107	11449	30	900	3210
171	29241	24	576	4104
250	62500	20	400	5000
171	29241	27	729	4617
250	62500	20	400	5000
231	53361	26	676	6006
225	50625	23	529	5175
231	53361	26	676	6006
260	67600	21	441	5460
97	9409	35	1225	3395
156	24336	22	484	3422
240	57600	27	729	6480
111	12321	32	1024	3552

$\Sigma X = 3761$ $\Sigma X^2 = 789,475$ $\Sigma Y = 533$ $\Sigma Y^2 = 14,899$ $\Sigma XY = 93,810$

$\bar{X} = 188.05$ $\bar{Y} = 26.65$

4. Find the sum of the cross products for the data shown in the following distribution of paired scores.

(1) X	(2) X^2	(3) Y
4	16	25
5	25	23
5	25	23
6	36	21
6	36	21
6	36	21
7	49	19
7	49	19
8	64	17

$\Sigma X = 54$ $\Sigma X^2 = 336$ $\Sigma Y = 189$

(4) Y^2	(5) XY
625	
529	
529	
441	
441	
441	
361	
361	
289	

$\Sigma Y^2 = 4017$ $\Sigma XY =$

$\Sigma XY =$

$\dfrac{(\Sigma X)(\Sigma Y)}{N} =$

$\Sigma xy = \Sigma XY - \dfrac{(\Sigma X)(\Sigma Y)}{N}$

$=$

$=$

Since Σxy is negative, the correlation coefficient is negative.

A. $\Sigma xy = \Sigma XY - \dfrac{(\Sigma X)(\Sigma Y)}{N}$

$\qquad = 93{,}810 - 100{,}230.65$

$\qquad = -6420.65$

B. $\Sigma x^2 = \Sigma X^2 - \dfrac{(\Sigma X)^2}{N}$

$\qquad = 789{,}475 - 707{,}256.05$

$\qquad = 82{,}218.95$

C. $\Sigma y^2 = \Sigma Y^2 - \dfrac{(\Sigma Y)^2}{N}$

$\qquad = 14{,}899 - 14{,}204.45$

$\qquad = 694.55$

D. $r = \dfrac{-6420.65}{\sqrt{(82{,}218.95)(694.55)}}$

$\quad = \dfrac{-6420.65}{7556.80}$

$\quad = -0.85$

Do Exercises 5 and 6 at the right.

CORRELATION COEFFICIENT WITH ORDINAL SCALES— Spearman rho (r_{rho})

The Spearman r is employed when one scale constitutes ordinal measurement and the remaining scale is either ordinal or higher. However, prior to calculating r_{rho}, both scales must be expressed as ranks.

Converting Scores to Ranks

With much research data, one of the scales is inherently ordinal (e.g., observer rankings of leadership qualities, aggressiveness, or dominance) and the other scale is quantitative (e.g., the weight of the organism). This quantitative scale must be converted to ranks prior to calculating Spearman r.

Imagine a study that involves rating ten experimental animals along a dominance–submissiveness continuum. The animal judged highest in dominance receives a rank of 1 and the one judged lowest in dominance receives a rank of 10, with the remaining animals judged to occupy intermediate positions. We wish to determine if the weight of

5. Substitute Σx^2, Σy^2, and Σxy obtained from Exercises 1 through 3 in the formula for r and solve:

$$r = \dfrac{\Sigma xy}{\sqrt{\Sigma x^2 \cdot \Sigma y^2}}$$

$\quad =$

6. Find the correlation between cubic inches of engine displacement and highway miles per gallon for twenty 1976 automobiles equipped with automatic transmission.

X	X^2	Y	Y^2	XY
236		19		
304		20		
97		32		
231		24		
140		27		
85		35		
168		22		
107		25		
171		20		
250		21		
171		23		
250		21		
231		24		
260		18		
97		26		
156		18		
240		23		
111		28		
225		20		
231		24		

the animals is correlated with position in the dominance-submissiveness hierarchy. Since weight is a quantitative variable, it must be converted to ranks. We assign the rank of 1 to the heaviest animal, 2 to the next in weight, and so forth until all ten ranks have been assigned from the lightest to the heaviest animals.

Example

The following weights were obtained on the ten experimental animals.

Weight (grams)	Rank (weight)	Rank (dominance)
390	1	1
305	3	2
298	5	3
362	2	4
302	4	5
258	7	6
279	6	7
240	8	8
234	9	9
230	10	10

Do Exercises 7 through 10 at the right.

Calculating r_{rho}

Once both scales are expressed as ranks, we are ready to calculate r_{rho}. Spearman r is defined as

$$r_{rho} = 1 - \frac{6\Sigma D^2}{N(N^2 - 1)}.$$

Rank (weight)	Rank (dominance)	D	D^2
1	1	0	0
3	2	1	1
5	3	2	4
2	4	−2	4
4	5	−1	1
7	6	1	1
6	7	−1	1
8	8	0	0
9	9	0	0
10	10	0	0
		$\Sigma D = 0$	$\Sigma D^2 = 12$

Convert the following scores to ranks.

7. 120
    142
    110
     90
     67
    185
     54

8.   15
      5
      9
     10
     11

9.   87
     72
     64
     53
     19
      5
      2
      1

10. 115
      86
     119
      73
     136
      62
     105
     112
      75
      97
     104

A. Subtract each rank in dominance from its corresponding rank in weight and place in the D (difference) column.

B. As a check, sum the D column. It should equal zero. If it does not, an error has been made. Recheck both rankings and subtractions.

C. Square each D and place in the D^2 column.

D. Sum the D^2 column to obtain ΣD^2. In the present problem, $\Sigma D^2 = 12$.

E. Count the number of pairs of ranks to obtain N. In the present example, $N = 10$.

F. Substitute ΣD^2 and N in the formula for Spearman r.

$$r_{\text{rho}} = 1 - \frac{6\Sigma D^2}{N(N^2 - 1)}$$

$$= 1 - \frac{6(12)}{10(100 - 1)}$$

$$= 1 - \frac{72}{990}$$

$$= 1 - 0.07$$

$$= 0.93$$

There is a high positive correlation between weight of animal and position on a dominance–submissiveness hierarchy.

Do Exercises 11 through 14 at the right.

Converting to Ranks When There Are Tied Scores

When you are ranking a set of quantitative scores, you will often find ties. The procedure is to assign the mean rank to each of the tied scores.

Examples

Score	Rank
20	1
17	2.5
17	2.5
10	4

The two scores of 17 share the ranks 2 and 3. The mean is $\frac{2 + 3}{2} = 2.5$.

Find r_{rho} for the following sets of ranks.

11.

Rank on X	Rank on Y
1	8
3	7
2	6
4	5
7	4
6	3
5	2
8	1

12.

Rank on X	Rank on Y
1	1
2	2
3	3
4	4
5	5

13.

Rank on X	Rank on Y
1	3
2	1
3	6
4	5
5	2
6	4

14.

Rank on X	Rank on Y
12	1
8	2
10	3
9	4
4	5
6	6
1	7
2	8
7	9
11	10
3	11
5	12

Score	Rank
105	1
102	3
102	3
102	3
100	5

The three scores of 102 share the ranks of 2, 3, and 4. The mean is

$$\frac{2 + 3 + 4}{3} = 3.0.$$

Do Exercises 15 and 16 at the right.

Calculating Rank Coefficient when There Are Tied Scores

When there are tied ranks on either or both the X- and the Y-variables, the Spearman formula yields a spuriously high coefficient of correlation, particularly when the number of tied ranks is high. When there are ties, the Pearson r formula should be *applied to the ranked data.*

Example

Rank on X	X^2	Rank on Y	Y^2	XY
1	1	2	4	2
2	4	3	9	6
3.5	12.25	1	1	3.5
3.5	12.25	4.5	20.25	15.75
5	25	4.5	20.25	22.5
7	49	6	36	42
7	49	8	64	56
7	49	7	49	49
9	81	9	81	81
$\Sigma X = 45$	$\Sigma X^2 = 282.5$	$\Sigma Y = 45$	$\Sigma Y^2 = 284.5$	$\Sigma XY = 277.75$

$$r = \frac{\Sigma xy}{\sqrt{(\Sigma x^2)(\Sigma y^2)}}$$

A. Find the sum of the cross products.

$$\Sigma xy = \Sigma XY - \frac{(\Sigma X)(\Sigma Y)}{N}$$

$$= 277.75 - 225$$

$$= 52.75$$

Convert the following sets of scores to ranks.

15.

Score	Rank
18	
12	
10	
11	
14	
19	
14	
4	
3	
14	

16.

Score	Rank
108	
101	
115	
108	
115	
103	

B. Find the sum of squares of the X ranks.

$$\Sigma x^2 = (\Sigma X)^2 - \frac{(\Sigma X)^2}{N}$$

$$= 282.5 - \frac{(45)^2}{9}$$

$$= 282.5 - 225$$

$$= 57.5$$

C. Find the sum of squares of the Y ranks.

$$\Sigma y^2 = \Sigma Y^2 - \frac{(\Sigma Y)^2}{N}$$

$$= 284.5 - 225$$

$$= 59.5$$

D. Substitute in the formula for Pearson r.

$$r = \frac{52.75}{\sqrt{(57.5)(59.5)}}$$

$$= \frac{52.75}{58.49}$$

$$= 0.90$$

Do Exercises 17 and 18 at the right.

Find the correlation for the following sets of scores where there are tied ranks.

17.

Rank on X	Rank on Y
1	1
3	2
3	3
3	4
5	5

18.

Rank on X	Rank on Y
1	1.5
2	1.5
3.5	3
3.5	6
5	5
6.5	4
6.5	7

CHAPTER 9 TEST

1. The accompanying scatter diagram shows:

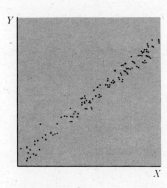

   a) a high positive correlation between X and Y

   b) little or no relationship between X and Y

   c) a high negative relationship between X and Y

   d) a low negative relationship between X and Y

2. If $\Sigma X = 87$, $\Sigma Y = 42$, $\Sigma XY = 202$, and $N = 15$, Σxy equals:

   a) 13.47   b) −41.6   c) 243.6   d) 3654

3. If $\Sigma X = 205$, $\Sigma Y = 94$, $\Sigma XY = 2306$, and $N = 9$, Σxy equals:

   a) 256.22   b) 2306   c) 164.89   d) 2141.11

4. If Σxy is zero, the correlation is:

   a) low negative

   b) high negative

   c) low positive

   d) none of the preceding

5. If r is negative, we know that:

   a) Σx^2 and Σxy are negative

   b) Σy^2 and Σxy are negative

   c) Σxy is negative

   d) either Σx^2 or Σy^2 is negative

6. Which of the following is a true statement?

   a) $r = -0.24$ is higher than $r = 0.00$

   b) $r = 0.86$ is higher than $r = -0.92$

   c) $r = -0.86$ is lower than $r = -0.45$

   d) $r = 0.64$ is lower than $r = -0.58$

7. Given $\Sigma xy = 16$, $\sqrt{\Sigma x^2 \cdot \Sigma y^2} = 81$, r equals:

   a) 0.56   b) 0.20   c) 0.05   d) 0.44

8. Given $\Sigma xy = 9.1$, $\sqrt{\Sigma x^2 \cdot \Sigma y^2} = 490$, r equals:

   a) 0.13   b) 0.41   c) 0.19   d) 0.02

9. In converting the scores 8, 14, 4, 2, 19, 12 to ranks (assigning rank 1 to the highest score), the score of 2 has a corresponding rank of:

   a) 2   b) 6   c) 1   d) 7

10. In converting the scores 6, 18, 12, 5, 9, 12, 1 to ranks (assigning rank 1 to the highest score), the score of 12 has a corresponding rank of:

    a) 2   b) 3   c) 2.5   d) 5.5

11. If a person ranks lowest on sociability and highest on introversion and another person scores highest on sociability and lowest on introversion, the rho coefficient is probably:

    a) zero   b) high positive

    c) low positive   d) high negative

12. If $\dfrac{6\Sigma D^2}{N(N^2 - 1)}$ is zero, r_{rho} is:

    a) 00.0   b) high negative

    c) +1.00   d) low positive

13. Given the pairs of ranks 4,2; 1,3; 2,1; 5,6; 6,5; 3,4, ΣD^2 is:

    a) 8   b) 64   c) 36   d) 12

14. Given: $\Sigma D^2 = 75$, $N = 10$, r_{rho} is:

    a) 0.45   b) 0.55   c) 0.08   d) 0.92

15. Given $\Sigma D^2 = 140$, $N = 11$, r_{rho} is:

    a) −0.36   b) 0.64   c) 0.11   d) 0.89

Practical

16. Following are the scores on an entrance exam of first-year college students and their grade point averages upon completion of the first semester. Prepare a scatter diagram and calculate Pearson *r* for these data.

Student	Entrance exam score	Grade point average
A	40	1.57
B	48	1.83
C	55	2.05
D	60	1.14
E	73	2.73
F	85	1.65
G	89	2.02
H	100	2.98
I	112	1.79
J	118	2.63
K	128	2.08
L	150	2.15
M	182	3.44
N	169	3.05
O	185	3.19
P	193	3.42
Q	220	3.87
R	250	3.00
S	290	3.12

17. A group of salespersons were rank-ordered in terms of assertiveness. Their gross sales for the year were subsequently tabulated. Determine if there is a correlation between assertiveness and sales performance.

Salesperson	Rank in assertiveness	Sales in thousands
A	1	273
B	2	215
C	3	244
D	4	265
E	5	205
F	6	230
G	7	195
H	8	220
I	9	163
J	10	157
K	11	178
L	12	145
M	13	158

Regression and Prediction

In this chapter we take up the equation for the best-fitting straight line for bivariate distributions and show how the linear relationship between two variables can be used: (1) to construct regression lines (prediction lines) relating one variable to the second; (2) to predict unknown values of one variable from known values of another variable.

THE BEST-FITTING STRAIGHT LINE FOR BIVARIATE (TWO-VARIABLE) DISTRIBUTIONS— PREDICTING *Y* FROM *X*

The best-fitting straight line for predicting *Y*-values from known values of *X* is

$Y'(Y \text{ predicted}) = a + bX.$

If we know the value of the two coefficients (*a* and *b*) in the above formula, we may predict values of *Y* from various values of the variable *X*.

Examples

Given: $a = 5$, $b = 1.3$.

When $X = 10$, $Y' = 5 + 1.3(10)$
$\qquad\qquad = 5 + 13 = 18.$

When $X = 0$, $Y' = 5 + 1.3(0)$
$\qquad\qquad = 5.$

When $X = 3$, $Y' = 5 + 1.3(3)$
$\qquad\qquad = 8.9.$

Do Exercises 1 through 5 at the right.

Constructing the Best-Fitting Straight Line

We may construct the best-fitting straight line by taking two extreme values of *X* (e.g., 0 and 10 in the example above), plotting the *Y'* for each of these values, and drawing a straight line between the two predicted values.

Example

When $a = 5$ and $b = 1.3$, predicted *Y* (*Y'*) is 5 when $X = 0$ and 18 when $X = 10$.

OBJECTIVES

- Know how to calculate the best-fitting straight line for predicting *Y*-values from known values of the *X*-variable.
- Know how to calculate the best-fitting straight line for predicting *X*-values from known values of the *Y*-variable.
- Using linear regression, know how to predict values of *Y* from known values of *X* and, conversely, values of *X* from known values of *Y*.
- Know how to calculate and interpret the standard error of estimate.

EXERCISES

Given $a = 2$, $b = 0.7$, find predicted *Y* (*Y'*) for the following values of *X*.

1. $X = 30$

2. $X = 1$

3. $X = 15$

4. $X = 9$

5. $X = 0$

10

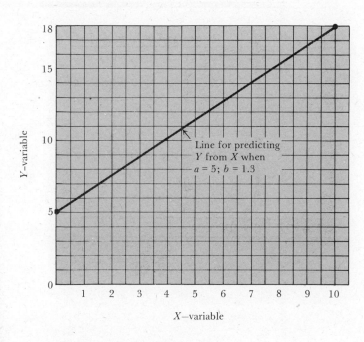

Do Exercise 6 at the right.

Finding the b_{yx} Coefficient

We shall use the data and prior calculations (Chapter 9) of cubic inches of engine displacement and highway miles-per-gallon scores of twenty 1976 automobiles. We found

$\Sigma xy = -6420.65$

$\Sigma x^2 = 82,218.95$

$\Sigma y^2 = 694.55$

$r = -0.85$

$N = 20$.

The equation for the b coefficient used in the prediction of Y from X is

$$b_{yx} = \frac{\Sigma xy}{\Sigma x^2} = \frac{-6420.65}{82,218.95} = -0.08.$$

The symbol b_{yx} reads: the b coefficient for the prediction of Y from X or, alternatively, the b coefficient for the regression of Y on X.

Do Exercises 7 through 10 at the right.

6. Plot the best-fitting straight line for the data in Exercises 1 through 5.

Given the values of Σxy and Σx^2, find the b coefficient for the regression of Y on X (b_{yx}).

7. $\Sigma xy = 20$; $\Sigma x^2 = 15$

8. $\Sigma xy = -10$; $\Sigma x^2 = 10$

9. $\Sigma xy = 28$; $\Sigma x^2 = 59$

10. $\Sigma xy = 0$; $\Sigma x^2 = 30$

Finding the a_{yx} Coefficient

The a coefficient is the value of Y when $X = 0$. It represents the value of Y where the regression line intercepts the Y-axis.

To find the a coefficient we must know the mean of X, the mean of Y, and the b coefficient for the regression of Y on X. Using the miles-per-gallon data, $\bar{X} = 188.05$; $\bar{Y} = 26.65$; and $b_{yx} = -0.08$.

The a coefficient of Y on X is

$$a_{yx} = \bar{Y} - b_{yx}(\bar{X})$$

$$= 26.65 - (-0.08)(188.05)$$

$$= 26.65 + 15.04$$

$$= 41.69.$$

Do Exercises 11 through 14 at the right.

The Equation of the Line for Predicting Y from X

As we saw previously, the line for predicting Y from X is

$$Y' = a_{yx} + b_{yx}(X).$$

We may now combine our calculations of b_{yx} and a_{yx} to write the equation of the regression line for predicting miles per gallon from knowing cubic inches of engine displacement.

$$Y' = 41.69 + (-0.08)X$$

$$= 41.69 - 0.08X$$

Do Exercises 15 through 18 at the right.

Using the Regression Equation to Predict Y from Known Values of X

Once we have obtained the equation for the regression of Y on X, we may use this equation to predict Y-values (miles per gallon) from knowledge of X (cubic inches of engine displacement).

Given the following values of $\bar{X}$, $\bar{Y}$, and b_{yx}, find a_{yx}.

11. $\bar{X} = 5$, $\bar{Y} = 20$, $b_{yx} = 1.33$

12. $\bar{X} = 4.5$, $\bar{Y} = 19.4$, $b_{yx} = -1.00$

13. $\bar{X} = 8.7$, $\bar{Y} = 142$, $b_{yx} = 0.47$

14. $\bar{X} = 65.4$, $\bar{Y} = 101.3$, $b_{yx} = 0.00$

Given b_{yx} and a_{yx}, write the equation for the regression line of Y on X.

15. $a_{yx} = 13.35$; $b_{yx} = 1.33$

16. $a_{yx} = 23.90$; $b_{yx} = -1.00$

17. $a_{yx} = 137.91$; $b_{yx} = 0.47$

18. $a_{yx} = 65.40$; $b_{yx} = 0.00$

10

Example

Imagine we have two cars for which mileage ratings are not available. We know that the cubic inches of engine displacement of Car A is 90 ($X_A = 90$) and Car B is 300 ($X_B = 300$). What are the predicted miles per gallon of each car?

$$Y' = a_{yx} + b_{yx}(X)$$

$$Y'_A = 41.69 - 0.08(90)$$

$$= 41.69 - 7.20$$

$$= 34.49$$

$$Y'_B = 41.69 - 0.08(300)$$

$$= 17.69$$

Do Exercises 19 through 22 at the right.

Constructing the Regression Line of Y on X (the line for predicting Y from known values of X)

When you have calculated the regression line of Y on X, you may superimpose it on the scatter diagram: calculate the predicted Y for two extreme values of X, plot these predicted scores on the scatter diagram, and then join the two points by a straight line.

(As a check on the accuracy of the calculations and/or drawings, it is helpful to know that the regression line *always* passes through the point where the means of X and Y intersect.)

Example

Using the cubic inches of engine displacement and the miles-per-gallon data, we have previously calculated that

when $X = 90$, $Y' = 34.49$,

when $X = 300$, $Y' = 17.69$.

We plot these points on the scatter diagram and connect them with a straight line to obtain the regression line for predicting Y-values from X-values.

Using the regression equation shown, predict the Y-scores for the indicated X-values.

19. $Y' = 13.35 + 1.33(X)$
   a) When $X = 10$
   b) When $X = 40$

20. $Y' = 23.90 - 1.00(X)$
   a) When $X = 20$
   b) When $X = 5$

21. $Y' = 66.31 + 0.47(X)$
   a) When $X = 80$
   b) When $X = 140$

22. $Y' = 65.40 + 0(X)$
   a) When $X = 65$
   b) When $X = 90$

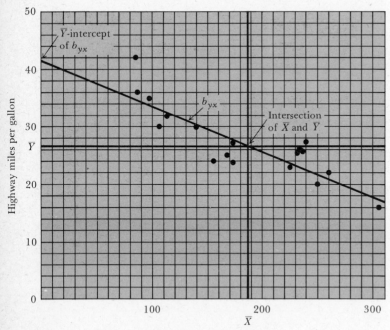

Cubic inches of engine displacement

Do Exercises 23 through 26 at the right.

THE BEST-FITTING STRAIGHT LINE FOR BIVARIATE DISTRIBUTIONS—PREDICTING X FROM Y

Except for the special case in which $r = \pm 1.00$, there are two separate regression lines that can be fitted to a bivariate distribution. One is for predicting Y from X and the other is for predicting X from Y. The procedures are identical, although the formulas are somewhat different.

The best-fitting straight line for predicting X-values from known values of Y is

$$X' = a_{xy} + bY_{xy}.$$

Example

If $a_{xy} = 16$, $b_{xy} = -0.7$, and $Y = 15$,

$$X' = 16 + (-0.7)(15)$$
$$= 16 + (-10.5) = 5.5.$$

Do Exercise 27 at the right.

Draw the regression line for the preceding exercises.

23. Exercise 19

24. Exercise 20

25. Exercise 21

26. Exercise 22

27. Given $a = 16$ and $b = -0.7$, find X' when
    a) $Y = 25$,
    b) $Y = 42$.

10

Finding the b_{yx} Coefficient

We shall use the data or prior calculations (Chapter 9) of cubic inches of engine displacement and highway miles-per-gallon scores of twenty 1976 automobiles. We found

$\Sigma xy = -6420.65$

$\Sigma x^2 = 82{,}218.95$

$\Sigma y^2 = 694.55$

$r = -0.85$

$N = 20.$

The equation of the b coefficient used in the prediction of X from Y is

$$b_{xy} = \frac{\Sigma xy}{\Sigma y^2}$$

$$= \frac{-6420.65}{694.55}$$

$$= -9.24.$$

[Note: We have now calculated two b coefficients; b_{yx} and b_{xy}. As a check on accuracy of calculations, the product of these two coefficients should equal r^2, allowing for some rounding error. In the present example in which $r = -0.85$ and $r^2 = 0.72$,

$b_{yx} = -0.08$ and $b_{xy} = -9.24.$

Thus,

$(b_{yx})(b_{xy}) = 0.74.$

The discrepancy of 0.02 represents rounding error.]

Do Exercises 28 through 30 at the right.

Finding the a_{xy} Coefficient

The a coefficient is the value of X when $Y = 0$. It represents the value of X when the regression line intercepts the X-axis.

To find the a_{xy} coefficient, we must know $\bar{X}$, $\bar{Y}$, and the b coefficient for the regression of X on Y. From the miles-per-gallon data,

$\bar{X} = 188.05$

$\bar{Y} = 26.65$

$b_{xy} = -9.24.$

Given the following values of Σxy and Σy^2, find the b coefficient for the regression of X on Y (b_{xy}).

28. $\Sigma xy = 20$; $\Sigma y^2 = 74$

29. $\Sigma xy = -10$; $\Sigma y^2 = 15.62$

30. $\Sigma xy = 28$; $\Sigma y^2 = 82.35$

The *a* coefficient of *X* on *Y* is

$$a_{xy} = \bar{X} - b_{xy}(\bar{Y})$$

$$= 188.05 - (-9.24)(26.65)$$

$$= 434.30.$$

Do Exercises 31 through 33 at the right.

The Equation of the Line for Predicting *X* from *Y*

As we noted previously, the line for predicting *X* from *Y* is

$$X' = a_{xy} + b_{xy}(Y).$$

We may now combine our calculations of b_{xy} and a_{xy} to write the regression line for "predicting" cubic inches of engine displacement from knowledge of miles per gallon.

$$X' = 434.30 + (-9.24)(Y)$$

Do Exercises 34 through 36 at the right.

Using the Regression Equation to Predict *X* from Known Values of *Y*

Once we have obtained the equation for the regression of *X* on *Y*, we may use this equation to "predict" *X*-values (cubic-inch displacement) from knowledge of *Y* (miles per gallon).

Example

Imagine we have two cars for which mileage ratings are available but we do not know the cubic inches of engine displacement. We know that the miles per gallon for Car A are 34.49 ($Y_A = 34.49$) and for Car B, 17.69 ($Y_B = 17.69$). What are the "predicted" cubic inches of displacement of each car?

$$X' = a_{xy} + b_{xy}(Y)$$

$$X'_A = 434.30 + (-9.24)(34.49)$$

$$= 434.40 - 318.69$$

$$= 115.61$$

$$X'_B = 434.30 + (-9.24)(17.69)$$

$$= 434.30 - 163.46$$

$$= 270.84$$

Do Exercises 37 through 39 at the right.

Given the following values of $\bar{X}$, $\bar{Y}$, and b_{xy}, find a_{xy}.

31. $\bar{X} = 5$; $\bar{Y} = 20$; $b_{xy} = 0.27$

32. $\bar{X} = 4.5$; $\bar{Y} = 19.4$; $b_{xy} = -0.64$

33. $\bar{X} = 8.7$; $\bar{Y} = 142$; $b_{xy} = 0.34$

Given b_{xy} and a_{xy}, write the equation for the regression line of *X* on *Y*.

34. $a_{xy} = -0.04$; $b_{xy} = 0.27$

35. $a_{xy} = 16.92$; $b_{xy} = -0.64$

36. $a_{xy} = -39.58$; $b_{xy} = 0.34$

Using the regression equation shown, predict the *X*-score for the indicated *Y*-values.

37. $X' = -0.4 + 0.27(Y)$
   a) When $Y = 12$
   b) When $Y = 28$

38. $X' = 16.92 - 0.64(Y)$
   a) When $Y = 8$
   b) When $Y = 30$

39. $X' = -39.58 + 0.34(Y)$
   a) When $Y = 100$
   b) When $Y = 180$

10

Constructing the Regression Line of X on Y (the line for predicting X from known values of Y)

When you have calculated the regression line of X on Y, you may superimpose it on the scatter diagram: calculate predicted X for two extreme values of Y, plot these predicted scores on the scatter diagram, and then join the two points by a straight line.

(Recall that the regression line will pass through the intersection of $\overline{X}$ and $\overline{Y}$).

Example

Using the miles-per-gallon data and the cubic inches of engine displacement, we have previously calculated that

when $Y = 34.49$, $X' = 115.61$,

when $Y = 17.69$, $X' = 270.84$.

We plot these points on the scatter diagram and connect them with a straight line to obtain the regression line for predicting X-values from Y-values.

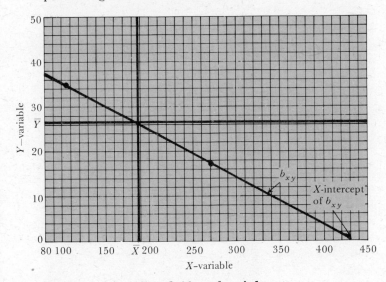

Do Exercises 40 and 41 at the right.

Draw the regression line for the following exercises.

40. Exercise 37.

41. Exercise 38.

An Alternate Method for Obtaining X' and Y'

If you have information about the standard deviations of X and Y, the means of X and Y, and the correlation between X and Y, alternate formulas are available that do not require the calculation of a and b. These formulas are

$$X' = \bar{X} + r \frac{sx}{sy} (Y - \bar{Y})$$

and

$$Y' = \bar{Y} + r \frac{sy}{sx} (X - \bar{X}).$$

Example

Given:

$\bar{X} = 188.05$

$s_x = 65.78$

$Y = 26.65$

$s_y = 6.05$

$r = -0.85$

find Y' when $X = 300$.

$$Y' = 26.65 + (-0.85) \frac{6.05}{65.78} (300 - 188.05)$$

$$= 26.65 - \frac{575.70}{65.78}$$

$$= 26.65 - 8.75 = 17.90*$$

Do Exercise 42 at the right.

THE STANDARD ERROR OF ESTIMATE OF X AND Y

In Chapters 6 and 7, we learned about the calculation and interpretation of an important measure of dispersion—the standard deviation. We saw that the standard deviation is based upon the deviations of scores from the mean of a distribution. These are then squared and summed and divided by $N - 1$.† The square root of the resulting quantity is the standard deviation.

---

* The slight disparity between the answer given here and the one obtained earlier represents rounding error.

† Recall that some authors prefer to use N in the denominator when calculating the standard deviation of a sample.

42. Given:

$\bar{X} = 188.05$

$s_x = 65.78$

$\bar{Y} = 26.65$

$s_y = 6.05$

$r = -0.85$

Find X' when $Y = 34.49$.

10

There is a measure of dispersion around each regression line that is analogous to the standard deviation. It is called the standard error of estimate. By taking each score and subtracting from it the predicted score (i.e., $X - X'$), we would have a deviation analogous to a deviation of a score from the mean $(X - \bar{X})$. The sum of the square of these deviations (i.e., $\Sigma(X - X')^2$) from the regression line would be analogous to the sum of squares about the mean of X (i.e., $\Sigma(X - \bar{X})^2$). Thus, the standard error of estimate of X may be defined as follows:

$$s_{\text{est } X} = \sqrt{\frac{\Sigma(X - X')^2}{N - 1}}$$

Similarly the squared deviations about the regression line of Y on X (i.e., the regression line for predicting Y-values from X) yields the standard error of estimate of Y:

$$s_{\text{est } Y} = \sqrt{\frac{\Sigma(Y - Y')^2}{N - 1}}$$

Fortunately there is a simplified formula that requires only that you know the correlation coefficient and the respective standard deviations of X and Y.

The Standard Error of Estimate of X

The standard error of estimate of X may be defined as follows:

$$s_{\text{est } X} = s_x \sqrt{1 - r^2}$$

Examples

Given: $r = 0.90$, $s_x = 6.42$.

Refer to Table A-2 (column 7) in the Appendix for $\sqrt{1 - r^2}$ corresponding to $r = 0.90$ (column 1). Here we find that when $r = 0.90$, $\sqrt{1 - r^2} = 0.4359$.

$$s_{\text{est } X} = (6.42)(0.4359)$$

$$= 2.80$$

Given: $r = 0.70$, $s_x = 6.42$.

Referring to Table A-2, we find $\sqrt{1 - r^2}$ corresponding to $r = 0.70$ equals 0.7141.

$$s_{\text{est } X} = (6.42)(0.7141)$$

$$= 4.58$$

Given: $r = 0.30$, $s_x = 6.42$.

Referring to Table A-2, we find $\sqrt{1 - r^2}$ corresponding to $r = 0.30$ equals 0.9539.

$s_{\text{est } X} = (6.42)(0.9539)$

$\qquad = 6.12$

Note that the larger the magnitude of r, the smaller the standard error of estimate. This means that there is less dispersion of scores about the regression line as r increases. In the limiting case, when $r = 1.00$ and $\sqrt{1 - r^2}$ equals 0.00, the standard error of estimate is zero—all data points fall on the regression line so that all deviations are equal to zero.

Do Exercises 43 through 53 at the right.

The Standard Error of Estimate of Y

The standard error of estimate of Y is defined as follows:

$s_{\text{est } Y} = s_y \sqrt{1 - r^2}$

Example

Given: $r = 0.45$, $s_y = 16.93$.

Referring to Table A-2 (column 7) for $\sqrt{1 - r^2}$ corresponding to $r = 0.45$ (column 1), we find that when $r = 0.45$, $\sqrt{1 - r^2} = 0.8930$.

$s_{\text{est } Y} = (16.93)(0.8930)$

$\qquad = 15.12$

Do Exercises 54 through 57 at the right.

THE INTERPRETATION OF THE STANDARD ERROR OF ESTIMATE

We previously noted that the regression line in bivariate distributions is analogous to the mean in single variable distributions, and the standard error of estimate is analogous to the standard deviation. Recall that, for normally distributed variables, knowledge of the mean and standard deviation permits a precise interpretation of every score in a distribution with respect to percentile rank, percentage

Given the coefficients of correlation and the standard deviation of X, find $s_{\text{est } X}$ for the following:

43. $r = 1.00$; $s_x = 12.83$

44. $r = 0.85$; $s_x = 12.83$

45. $r = 0.65$; $s_x = 12.83$

46. $r = 0.50$; $s_x = 12.83$

47. $r = 0.20$; $s_x = 12.83$

48. $r = 0$; $s_x = 12.83$

49. $r = -0.15$; $s_x = 12.83$

50. $r = -0.30$; $s_x = 12.83$

51. $r = -0.70$; $s_x = 12.83$

52. $r = -0.95$; $s_x = 12.83$

53. $r = -1.00$; $s_x = 12.83$

Given the coefficients of correlation and the standard deviation of Y, find $s_{\text{est } Y}$ for each of the following:

54. $r = 0.95$; $s_y = 0.92$

55. $r = 0.75$; $s_y = 5.63$

56. $r = 0.10$; $s_y = 20$

57. $r = -0.95$; $s_y = 0.92$

10

of area above and below a given score, and the percentage of area between the mean and a given score.

If variable X is normally distributed at all values of Y and Y is normally distributed at all values of X (a condition known as *homoscedasticity*), precise interpretations of scores are possible with respect to the regression line (predicted scores).

Finding and Interpreting z_{xy}

Recall that, except for the special case in which $r = \pm 1.00$ (when all scores fall precisely on the regression line), the scores are distributed or dispersed around the regression line for *all* values of r less than perfect. The higher the correlation, the less the dispersion; the lower the correlation, the greater the dispersion. It is possible to obtain a z-score for any predicted value of X and interpret this score with respect to the standard normal distribution.

Example

Given: $X = 12.10$, $X' = 10.63$, $s_{\text{est }X} = 2.62$.

Find z_{xy}.

$$z_{xy} = \frac{X - X'}{s_{\text{est }X}}$$

$$= \frac{12.10 - 10.63}{2.62}$$

$$= \frac{1.47}{2.62}$$

$$= 0.56$$

Do Exercises 58 through 60 at the right.

The z-score may now be interpreted by use of Table A-1, which shows areas under the standard normal curve.

The $z_{xy} = 0.56$ in the above example means that the score of 12 was 0.56 standard errors of estimate above the predicted value. The percent of area between the predicted score, X', and the obtained score, X, is found to be 21.23. The percentile rank of the score, among those for whom the predicted value was 10.63, is found (Table A-1, column A) to be 71.23.

Do Exercises 61 through 63 at the right.

Given the following values of X, X', and $s_{\text{est }X}$, find z_{xy}.

58. $X = 115$; $X' = 105$; $s_{\text{est }X} = 6$

59. $X = 80$; $X' = 97$; $s_{\text{est }X} = 8$

60. $X = 103$; $X' = 100$; $s_{\text{est }X} = 12$

Given the following values of z_{xy}, find the corresponding percentile rank.

61. $z_{xy} = -1.23$

62. $z_{xy} = 1.65$

63. $z_{xy} = -0.62$

Finding and Interpreting z_{yx}

All scores fall directly on the regression line only when $r = \pm 1.00$. At all other times, the obtained scores are dispersed about the regression line. The extent of variability about the regression line depends upon the magnitude of r. When r is low, the dispersion—as represented by the standard error of estimate—is large. When r is high, the dispersion is small.

It is possible to obtain a z-score for any predicted value of X and interpret this score with respect to the standard normal curve.

Example

Given: $Y' = 16.53$, $Y = 14.92$, $s_{est\ Y} = 1.75$.

Find the percentage of cases, for which $Y' = 16.53$, that obtained scores of 14.92 or lower.

$$\frac{14.92 - 16.53}{1.75} = \frac{-1.61}{1.75}$$
$$= -0.92$$

Do Exercises 64 through 66 at the right.

The z_{yx} of -0.92 in the above example means that the obtained score of 14.92 was 0.92 standard error of estimate below the predicted score. The percentile rank of the score, among those for whom the predicted value was 16.53, is found (Table A-1, column A) to be 17.88. Stated another way: Among those whose predicted score was 16.53, 82.12 percent scored higher than 14.92.

Do Exercises 67 through 69 at the right.

A Practical Application of Regression Analysis in Education

Miss Carter is a fifth-grade teacher in a school district in which most children come from affluent homes. She has taken a special interest in one ten-year-old boy—Johnny G. In the classroom situation, Johnny has shown himself to be witty, resourceful, independent, and at times, extremely trying to the teacher's patience. Her general impression is that Johnny is extremely bright. However, his classroom performance appears to belie this general impression. Miss Carter has concluded that Johnny is working well below his level of capability. In short, he is an underachiever.

Given the following values of Y, Y', and $s_{est\ Y}$, find z_{yx}.

64. $Y = 42$; $Y' = 60$; $s_{est\ Y} = 21.40$

65. $Y = 138$; $Y' = 100$; $s_{est\ Y} = 16.42$

66. $Y = 8.63$; $Y' = 8.93$; $s_{est\ Y} = 2.46$

Given the following values of z_{yx}, find the corresponding percentile rank.

67. $z = -0.84$

68. $z = 2.31$

69. $z = -0.12$

Searching through the school records, she compiles the following information on Johnny. (For our purposes, we shall assume that the basic assumption of homoscedasticity is met.)

Score on test X— (academic aptitude) is 135	Grade point average
$\bar{X} = 100$	$Y = 1.62$
$s_x = 15$	$\bar{Y} = 1.56$
$r_{xy} = 0.70$	$s_Y = 0.80$

A. Find Y' for Johnny. In the present example,

$$Y' = \bar{Y} + r\left(\frac{s_y}{s_x}\right)(X - \bar{X})$$

$$= 1.56 + 0.70 \left(\frac{0.80}{15}\right)(135 - 100)$$

$$= 1.56 + 1.31$$

$$= 2.87.$$

Johnny's predicted grade point average is 2.87.

B. Find $s_{\text{est } Y}$. In the present example,

$$s_{\text{est } Y} = s_y\sqrt{1 - r^2}$$

$$= (0.80)(0.7141)$$

$$= 0.57.$$

C. Subtract predicted Y from obtained Y. In the present illustration,

$$Y - Y' = 1.62 - 2.87 = -1.25.$$

D. Find z_{yx}. In the present case,

$$z_{yx} = \frac{-1.25}{0.57} = -2.19.$$

E. Find the percentile rank of the z calculated in Step D (Table A-1, column A).

Referring to the present example, the percentile rank of $z = -2.19$ is 1.43.

Conclusion: Miss Carter's observation appears warranted. Among fifth graders who scored 135 on X, fully 98.57 percent obtained higher grade point averages than Johnny.

Do Exercise 70 at the right.

70. Margaret O. obtained the following scores on the test for academic aptitude (variable X) and grade point average (variable Y)

Variable X	Variable Y
$X = 115$	$Y = 3.51$
$\bar{X} = 100$	$\bar{Y} = 1.56$
$s_x = 15$	$s_y = 0.80$
$r = 0.70$	

Determine how well or how poorly she is performing relative to what she would be expected to perform.

CHAPTER 10 TEST

Use the data given in each problem and Tables A-1 and A-2 in the Appendix to complete this test.

1. Given the equation of the straight line, $Y' = a + bX$, and the values of $a = 15$, $b = -10$, and $X = 3$, the value of Y' is:

   a) 22   b) 35   c) 15   d) 45

2. Given $b_{yx} = \dfrac{\Sigma xy}{\Sigma x^2}$ and $\Sigma xy = -12.50$, $\Sigma x^2 = 32.10$, b_{yx} equals:

   a) -0.39 b) -2.57 c) 0.39   d) 2.57

3. Given $a_{yx} = \bar{Y} - b_{yx}(\bar{X})$ and $\bar{Y} = 1.87$, $b_{yx} = 0.25$ and $\bar{X} = 12.45$, a_{yx} equals:

   a) 9.34   b) 10.33   c) 11.98   d) -1.24

4. In the equation $Y' = 109.73 + 1.58X$, if $X = 100$, Y' equals:

   a) 125.53   b) 211.31   c) 267.73   d) -48.27

5. In the equation $X' = 0.54 + (-0.73)Y$, if $Y = 0.80$, X' equals:

   a) 1.12   b) -0.19 c) -0.04 d) 0.61

6. In the equation $X' = \bar{X} + r\dfrac{s_x}{s_y}(Y - \bar{Y})$, if $Y = 192.50$,

   $\bar{X} = 15.65$, $r = 0.55$, $s_x = 4.3$, $s_y = 25$, and $\bar{Y} = 170$, X' equals:

   a) 87.60   b) 17.78   c) 13.52   d) 186.73

7. In the equation $Y' = \bar{Y} + r\dfrac{s_y}{s_x}(X - \bar{X})$, if $\bar{Y} = 18.0$, $r = -0.95$,

   $s_x = 5$, $s_y = 3.2$, $\bar{X} = 30$, and $X = 20$, Y' equals:

   a) 32.84   b) 11.92   c) 3.16   d) 24.08

8. Given $s_{\text{est } X} = s_x\sqrt{1 - r^2}$ and $s_x = 4.53$, $s_y = 8.26$ and $r = 0.24$, $s_{\text{est } X}$ equals:

   a) 7.88   b) 8.02   c) 4.39   d) 4.27

9. Given $s_{\text{est } Y} = s_y\sqrt{1 - r^2}$ and $s_y = 0.76$, $s_x = 8.93$, and $r = -1.00$, $s_{\text{est } Y}$ equals:

   a) 0.76   b) 0.00   c) 8.93   d) -0.76

10. Given $X = 16$, $X' = 12$, $r = 0.62$ and $s_{\text{est } X} = 2.93$, z_{xy} equals:

    a) 1.37   b) -1.74 c) 1.74   d) -1.37

11. Given $Y = 157.08$, $Y' = 169.54$, $r = 0.73$, $s_{\text{est } Y} = 15.92$, z_{yx} equals:

    a) -0.78 b) 1.25   c) -1.25 d) 0.78

12. Given $X = 25.4$, $Y = 38.9$, $X' = 33.5$, $Y' = 32.8$, $s_{\text{est } X} = 4.70$, $s_{\text{est } Y} = 5.02$, z_{yx} equals:

    a) 1.30   b) 1.22   c) -2.69 d) 2.69

10

13. Given $X' = 6.87$, $s_{\text{est } X} = 1.34$, and $s_{\text{est } Y} = 2.05$. Assuming that homoscedasticity is met, the percentile rank of $X = 5.93$ is:

    a) 24.20   b) 67.72   c) 75.80   d) 32.28

14. Given $Y' = 14.40$, $s_{\text{est } Y} = 2.63$, and $s_{\text{est } X} = 1.59$. Assuming that homoscedasticity is met, the percentile rank of $Y = 15.95$ is:

    a) 72.24   b) 16.60   c) 83.40   d) 27.76

15. Given $Y' = 42.15$, $s_Y = 8.96$ and $r = 0.56$. Assuming that homoscedasticity is met, the percentile rank of $Y = 51.63$ is:

    a) 10.03   b) 93.82   c) 6.18   d) 89.97

Practical

16. Given $\bar{X} = 5.92$, $\bar{Y} = 50.00$, $s_x = 1.5$, $s_y = 10.00$, $r_{xy} = 0.65$. Draw the regression lines of X on Y and Y on X.

17. Given $\Sigma x^2 = 15.50$, $\Sigma y^2 = 33.46$, and $\Sigma xy = -18.59$; $\bar{X} = 18.9$, $\bar{Y} = 26.4$.

    a) Find b_{yx} and b_{xy}.

    b) Find a_{yx} and a_{xy}.

18. Find $s_{\text{est } X}$ and $s_{\text{est } Y}$ in each of the following examples.

    a) $s_x = 48.53$, $s_y = 3.96$, $r = -0.93$

    b) $s_x = 12.59$, $s_y = 21.65$, $r = 0.80$

    c) $s_x = 103.6$, $s_y = 29.53$, $r = -0.47$

    d) $s_x = 8.62$, $s_y = 4.04$, $r = 0.23$

19. Assume that homoscedasticity has been met in this problem.

G. John obtained a score of 480 on a college entrance test in which the mean was 500 and the standard deviation 100. At the end of his sophomore year in college, he had achieved a grade point average of 3.95 in which the mean of his class was 2.88 and the standard deviation was 0.79. The correlation between the two variables is 0.60. How was G. John performing relative to his expected level of performance?

Appendix

Table A-1   Percent of area under the standard normal curve, percentile rank corresponding to a given z, and height of ordinate at z

z	(A) Percentile rank (percent of area below z)	(B) Percent of area from mean to z	(C) Percent of area beyond z	(D) Height of ordinate at z
−3.70	00.01	49.99	00.01	.0004
−3.60	00.02	49.98	00.02	.0006
−3.50	00.02	49.98	00.02	.0009
−3.40	00.03	49.97	00.03	.0012
−3.30	00.05	49.95	00.05	.0017
−3.25	00.06	49.95	00.06	.0020
−3.24	00.06	49.94	00.06	.0021
−3.23	00.06	49.94	00.06	.0022
−3.22	00.06	49.94	00.06	.0022
−3.21	00.07	49.93	00.07	.0023
−3.20	00.07	49.93	00.07	.0024
−3.19	00.07	49.93	00.07	.0025
−3.18	00.07	49.93	00.07	.0025
−3.17	00.08	49.92	00.08	.0026
−3.16	00.08	49.92	00.08	.0027
−3.15	00.08	49.92	00.08	.0028
−3.14	00.08	49.92	00.08	.0029
−3.13	00.09	49.91	00.09	.0030
−3.12	00.09	49.91	00.09	.0031
−3.11	00.09	49.91	00.09	.0032
−3.10	00.10	49.90	00.10	.0033
−3.09	00.10	49.90	00.10	.0034
−3.08	00.10	49.90	00.10	.0035
−3.07	00.11	49.89	00.11	.0036
−3.06	00.11	49.89	00.11	.0037
−3.05	00.11	49.89	00.11	.0038
−3.04	00.12	49.88	00.12	.0039
−3.03	00.12	49.88	00.12	.0040
−3.02	00.13	49.87	00.13	.0042
−3.01	00.13	49.87	00.13	.0043
−3.00	00.13	49.87	00.13	.0044
−2.99	00.14	49.86	00.14	.0046
−2.98	00.14	49.86	00.14	.0047
−2.97	00.15	49.85	00.15	.0048
−2.96	00.15	49.85	00.15	.0050
−2.95	00.16	49.84	00.16	.0051
−2.94	00.16	49.84	00.16	.0053
−2.93	00.17	49.83	00.17	.0055
−2.92	00.18	49.82	00.18	.0056
−2.91	00.18	49.82	00.18	.0058
−2.90	00.19	49.81	00.19	.0060
−2.89	00.19	49.81	00.19	.0061
−2.88	00.20	49.80	00.20	.0063
−2.87	00.21	49.79	00.21	.0065
−2.86	00.21	49.79	00.21	.0067
−2.85	00.22	49.78	00.22	.0069
−2.84	00.23	49.77	00.23	.0071
−2.83	00.23	49.77	00.23	.0073
−2.82	00.24	49.76	00.24	.0075
−2.81	00.25	49.75	00.25	.0077

Table A-1 *(Continued)*

z	(A) Percentile rank (percent of area below z)	(B) Percent of area from mean to z	(C) Percent of area beyond z	(D) Height of ordinate at z
−2.80	00.26	49.74	00.26	.0079
−2.79	00.26	49.74	00.26	.0081
−2.78	00.27	49.73	00.27	.0084
−2.77	00.28	49.72	00.28	.0086
−2.76	00.29	49.71	00.29	.0088
−2.75	00.30	49.70	00.30	.0091
−2.74	00.31	49.69	00.31	.0093
−2.73	00.32	49.68	00.32	.0096
−2.72	00.33	49.67	00.33	.0099
−2.71	00.34	49.66	00.34	.0101
−2.70	00.35	49.65	00.35	.0104
−2.69	00.36	49.64	00.36	.0107
−2.68	00.37	49.63	00.37	.0110
−2.67	00.38	49.62	00.38	.0113
−2.66	00.39	49.61	00.39	.0116
−2.65	00.40	49.60	00.40	.0119
−2.64	00.41	49.59	00.41	.0122
−2.63	00.43	49.57	00.43	.0126
−2.62	00.44	49.56	00.44	.0129
−2.61	00.45	49.55	00.45	.0132
−2.60	00.47	49.53	00.47	.0136
−2.59	00.48	49.52	00.48	.0139
−2.58	00.49	49.51	00.49	.0143
−2.57	00.51	49.49	00.51	.0147
−2.56	00.52	49.48	00.52	.0151
−2.55	00.54	49.46	00.54	.0154
−2.54	00.55	49.45	00.55	.0158
−2.53	00.57	49.43	00.57	.0163
−2.52	00.59	49.41	00.59	.0167
−2.51	00.60	49.40	00.60	.0171
−2.50	00.62	49.38	00.62	.0175
−2.49	00.64	49.36	00.64	.0180
−2.48	00.66	49.34	00.66	.0184
−2.47	00.68	49.32	00.68	.0189
−2.46	00.69	49.31	00.69	.0194
−2.45	00.71	49.29	00.71	.0198
−2.44	00.73	49.27	00.73	.0203
−2.43	00.75	49.25	00.75	.0208
−2.42	00.78	49.22	00.78	.0213
−2.41	00.80	49.20	00.80	.0219
−2.40	00.82	49.18	00.82	.0224
−2.39	00.84	49.16	00.84	.0229
−2.38	00.87	49.13	00.87	.0235
−2.37	00.89	49.11	00.89	.0241
−2.36	00.91	49.09	00.91	.0246
−2.35	00.94	49.06	00.94	.0252
−2.34	00.96	49.04	00.96	.0258
−2.33	00.99	49.01	00.99	.0264
−2.32	01.02	48.98	01.02	.0270
−2.31	01.04	48.96	01.04	.0277

Table A-1   *(Continued)*

z	(A) Percentile rank (percent of area below z)	(B) Percent of area from mean to z	(C) Percent of area beyond z	(D) Height of ordinate at z
−2.30	01.07	48.93	01.07	.0283
−2.29	01.10	48.90	01.10	.0290
−2.28	01.13	48.87	01.13	.0297
−2.27	01.16	48.84	01.16	.0303
−2.26	01.19	48.81	01.19	.0310
−2.25	01.22	48.78	01.22	.0317
−2.24	01.25	48.75	01.25	.0325
−2.23	01.29	48.71	01.29	.0332
−2.22	01.32	48.68	01.32	.0339
−2.21	01.36	48.64	01.36	.0347
−2.20	01.39	48.61	01.39	.0355
−2.19	01.43	48.57	01.43	.0363
−2.18	01.46	48.54	01.46	.0371
−2.17	01.50	48.50	01.50	.0379
−2.16	01.54	48.46	01.54	.0387
−2.15	01.58	48.42	01.58	.0396
−2.14	01.62	48.38	01.62	.0404
−2.13	01.66	48.34	01.66	.0413
−2.12	01.70	48.30	01.70	.0422
−2.11	01.74	48.26	01.74	.0431
−2.10	01.79	48.21	01.79	.0440
−2.09	01.83	48.17	01.83	.0449
−2.08	01.88	48.12	01.88	.0459
−2.07	01.92	48.08	01.92	.0468
−2.06	01.97	48.03	01.97	.0478
−2.05	02.02	47.98	02.02	.0488
−2.04	02.07	47.93	02.07	.0498
−2.03	02.12	47.88	02.12	.0508
−2.02	02.17	47.83	02.17	.0519
−2.01	02.22	47.78	02.22	.0529
−2.00	02.28	47.72	02.28	.0540
−1.99	02.33	47.67	02.33	.0551
−1.98	02.39	47.61	02.39	.0562
−1.97	02.44	47.56	02.44	.0573
−1.96	02.50	47.50	02.50	.0584
−1.95	02.56	47.44	02.56	.0596
−1.94	02.62	47.38	02.62	.0608
−1.93	02.68	47.32	02.68	.0620
−1.92	02.74	47.26	02.74	.0632
−1.91	02.81	47.19	02.81	.0644
−1.90	02.87	47.13	02.87	.0656
−1.89	02.94	47.06	02.94	.0669
−1.88	03.01	46.99	03.01	.0681
−1.87	03.07	46.93	03.07	.0694
−1.86	03.14	46.86	03.14	.0707
−1.85	03.22	46.78	03.22	.0721
−1.84	03.29	46.71	03.29	.0734
−1.83	03.36	46.64	03.36	.0748
−1.82	03.44	46.56	03.44	.0761
−1.81	03.51	46.49	03.51	.0775

Table A-1 *(Continued)*

z	(A) Percentile rank (percent of area below z)	(B) Percent of area from mean to z	(C) Percent of area beyond z	(D) Height of ordinate at z
−1.80	03.59	46.41	03.59	.0790
−1.79	03.67	46.33	03.67	.0804
−1.78	03.75	46.25	03.75	.0818
−1.77	03.84	46.16	03.84	.0833
−1.76	03.92	46.08	03.92	.0848
−1.75	04.01	45.99	04.01	.0863
−1.74	04.09	45.91	04.09	.0878
−1.73	04.18	45.82	04.18	.0893
−1.72	04.27	45.73	04.27	.0909
−1.71	04.36	45.64	04.36	.0925
−1.70	04.46	45.54	04.46	.0940
−1.69	04.55	45.45	04.55	.0957
−1.68	04.65	45.35	04.65	.0973
−1.67	04.75	45.25	04.75	.0989
−1.66	04.85	45.15	04.85	.1006
−1.65	04.95	45.05	04.95	.1023
−1.64	05.05	44.95	05.05	.1040
−1.63	05.16	44.84	05.16	.1057
−1.62	05.26	44.74	05.26	.1074
−1.61	05.37	44.63	05.37	.1092
−1.60	05.48	44.52	05.48	.1109
−1.59	05.59	44.41	05.59	.1127
−1.58	05.71	44.29	05.71	.1145
−1.57	05.82	44.18	05.82	.1163
−1.56	05.94	44.06	05.94	.1182
−1.55	06.06	43.94	06.06	.1200
−1.54	06.18	43.82	06.18	.1219
−1.53	06.30	43.70	06.30	.1238
−1.52	06.43	43.57	06.43	.1257
−1.51	06.55	43.45	06.55	.1276
−1.50	06.68	43.32	06.68	.1295
−1.49	06.81	43.19	06.81	.1315
−1.48	06.94	43.06	06.94	.1334
−1.47	07.08	42.92	07.08	.1354
−1.46	07.21	42.79	07.21	.1374
−1.45	07.35	42.65	07.35	.1394
−1.44	07.49	42.51	07.49	.1415
−1.43	07.64	42.36	07.64	.1435
−1.42	07.78	42.22	07.78	.1456
−1.41	07.93	42.07	07.93	.1476
−1.40	08.08	41.92	08.08	.1497
−1.39	08.23	41.77	08.23	.1518
−1.38	08.38	41.62	08.38	.1539
−1.37	08.53	41.47	08.53	.1561
−1.36	08.69	41.31	08.69	.1582
−1.35	08.85	41.15	08.85	.1604
−1.34	09.01	40.99	09.01	.1626
−1.33	09.18	40.82	09.18	.1647
−1.32	09.34	40.66	09.34	.1669
−1.31	09.51	40.49	09.51	.1691

Table A-1   *(Continued)*

z	(A) Percentile rank (percent of area below z)	(B) Percent of area from mean to z	(C) Percent of area beyond z	(D) Height of ordinate at z
−1.30	09.68	40.32	09.68	.1714
−1.29	09.85	40.15	09.85	.1736
−1.28	10.03	39.96	10.03	.1758
−1.27	10.20	39.80	10.20	.1781
−1.26	10.38	39.62	10.38	.1804
−1.25	10.56	39.44	10.56	.1826
−1.24	10.75	39.25	10.75	.1849
−1.23	10.93	39.07	10.93	.1872
−1.22	11.12	38.88	11.12	.1895
−1.21	11.31	38.69	11.31	.1919
−1.20	11.51	38.49	11.51	.1942
−1.19	11.70	38.30	11.70	.1965
−1.18	11.90	38.10	11.90	.1989
−1.17	12.10	37.90	12.10	.2012
−1.16	12.30	37.70	12.30	.2036
−1.15	12.51	37.49	12.51	.2059
−1.14	12.71	37.29	12.71	.2083
−1.13	12.92	37.08	12.92	.2107
−1.12	13.14	36.86	13.14	.2131
−1.11	13.35	36.65	13.35	.2155
−1.10	13.57	36.43	13.57	.2179
−1.09	13.79	36.21	13.79	.2203
−1.08	14.01	35.99	14.01	.2227
−1.07	14.23	35.77	14.23	.2251
−1.06	14.46	35.54	14.46	.2275
−1.05	14.69	35.31	14.69	.2299
−1.04	14.92	35.08	14.92	.2323
−1.03	15.15	34.85	15.15	.2347
−1.02	15.39	34.61	15.39	.2371
−1.01	15.62	34.38	15.62	.2396
−1.00	15.87	34.13	15.87	.2420
−0.99	16.11	33.89	16.11	.2444
−0.98	16.35	33.65	16.35	.2468
−0.97	16.60	33.40	16.60	.2492
−0.96	16.85	33.15	16.85	.2516
−0.95	17.11	32.89	17.11	.2541
−0.94	17.36	32.64	17.36	.2565
−0.93	17.62	32.38	17.62	.2589
−0.92	17.88	32.12	17.88	.2613
−0.91	18.14	31.86	18.14	.2637
−0.90	18.41	31.59	18.41	.2661
−0.89	18.67	31.33	18.67	.2685
−0.88	18.94	31.06	18.94	.2709
−0.87	19.22	30.78	19.22	.2732
−0.86	19.49	30.51	19.49	.2756
−0.85	19.77	30.23	19.77	.2780
−0.84	20.05	29.95	20.05	.2803
−0.83	20.33	29.67	20.33	.2827
−0.82	20.61	29.39	20.61	.2850
−0.81	20.90	29.10	20.90	.2874

Table A-1 *(Continued)*

z	(A) Percentile rank (percent of area below z)	(B) Percent of area from mean to z	(C) Percent of area beyond z	(D) Height of ordinate at z
−0.80	21.19	28.81	21.19	.2897
−0.79	21.48	28.52	21.48	.2920
−0.78	21.77	28.23	21.77	.2943
−0.77	22.06	27.94	22.06	.2966
−0.76	22.36	27.64	22.36	.2989
−0.75	22.66	27.34	22.66	.3011
−0.74	22.96	27.04	22.96	.3034
−0.73	23.27	26.73	23.27	.3056
−0.72	23.58	26.42	23.58	.3079
−0.71	23.89	26.11	23.89	.3101
−0.70	24.20	25.80	24.20	.3123
−0.69	24.51	25.49	24.51	.3144
−0.68	24.83	25.17	24.83	.3166
−0.67	25.14	24.86	25.14	.3187
−0.66	25.46	24.54	25.46	.3209
−0.65	25.78	24.22	25.78	.3230
−0.64	26.11	23.89	26.11	.3251
−0.63	26.43	23.56	26.43	.3271
−0.62	26.76	23.24	26.76	.3292
−0.61	27.09	22.91	27.09	.3312
−0.60	27.43	22.57	27.43	.3332
−0.59	27.76	22.24	27.76	.3352
−0.58	28.10	21.90	28.10	.3372
−0.57	28.43	21.57	28.43	.3391
−0.56	28.77	21.23	28.77	.3410
−0.55	29.12	20.88	29.12	.3429
−0.54	29.46	20.54	29.46	.3448
−0.53	29.81	20.19	29.81	.3467
−0.52	30.15	19.85	30.15	.3485
−0.51	30.50	19.50	30.50	.3503
−0.50	30.85	19.15	30.85	.3521
−0.49	31.21	18.79	31.21	.3538
−0.48	31.56	18.44	31.56	.3555
−0.47	31.92	18.08	31.92	.3572
−0.46	32.28	17.72	32.28	.3589
−0.45	32.64	17.36	32.64	.3605
−0.44	33.00	17.00	33.00	.3621
−0.43	33.36	16.64	33.36	.3637
−0.42	33.72	16.28	33.72	.3653
−0.41	34.09	15.91	34.09	.3668
−0.40	34.46	15.54	34.46	.3683
−0.39	34.83	15.17	34.83	.3697
−0.38	35.20	14.80	35.20	.3712
−0.37	35.57	14.43	35.57	.3725
−0.36	35.94	14.06	35.94	.3739
−0.35	36.32	13.68	36.32	.3752
−0.34	36.69	13.31	36.69	.3765
−0.33	37.07	12.93	37.07	.3778
−0.32	37.45	12.55	37.45	.3790
−0.31	37.83	12.17	37.83	.3802

Table A-1 *(Continued)*

z	(A) Percentile rank (percent of area below z)	(B) Percent of area from mean to z	(C) Percent of area beyond z	(D) Height of ordinate at z
−0.30	38.21	11.79	38.21	.3814
−0.29	38.59	11.41	38.59	.3825
−0.28	38.97	11.03	38.97	.3836
−0.27	39.36	10.64	39.36	.3847
−0.26	39.74	10.26	39.74	.3857
−0.25	40.13	09.87	40.13	.3867
−0.24	40.52	09.48	40.52	.3876
−0.23	40.90	09.10	40.90	.3885
−0.22	41.29	08.71	41.29	.3894
−0.21	41.68	08.32	41.68	.3902
−0.20	42.07	07.93	42.07	.3910
−0.19	42.47	07.53	42.47	.3918
−0.18	44.86	07.14	42.86	.3925
−0.17	43.25	06.75	43.25	.3932
−0.16	43.64	06.36	43.64	.3939
−0.15	44.04	05.96	44.04	.3945
−0.14	44.43	05.57	44.43	.3951
−0.13	44.83	05.17	44.83	.3956
−0.12	45.22	04.78	45.22	.3961
−0.11	45.62	04.38	45.62	.3965
−0.10	46.02	03.98	46.02	.3970
−0.09	46.41	03.59	46.41	.3973
−0.08	46.81	03.19	46.81	.3977
−0.07	47.21	02.79	47.21	.3980
−0.06	47.61	02.39	47.61	.3982
−0.05	48.01	01.99	48.01	.3984
−0.04	48.40	01.60	48.40	.3986
−0.03	48.80	01.20	48.80	.3988
−0.02	49.20	00.80	49.20	.3989
−0.01	49.60	00.40	49.60	.3989
0.00	50.00	00.00	50.00	.3989
0.01	50.40	00.40	49.60	.3989
0.02	50.80	00.80	49.20	.3989
0.03	51.20	01.20	48.80	.3988
0.04	51.60	01.60	48.40	.3986
0.05	51.99	01.99	48.01	.3984
0.06	52.39	02.39	47.61	.3982
0.07	52.79	02.79	47.21	.3980
0.08	53.19	03.19	46.81	.3977
0.09	53.59	03.59	46.41	.3973
0.10	53.98	03.98	46.02	.3970
0.11	54.38	04.38	45.62	.3965
0.12	54.78	04.78	45.22	.3961
0.13	55.17	05.17	44.83	.3956
0.14	55.57	05.57	44.43	.3951
0.15	55.96	05.96	44.04	.3945
0.16	56.36	06.36	43.64	.3939
0.17	56.75	06.75	43.25	.3932
0.18	57.14	07.14	42.86	.3925
0.19	57.53	07.53	42.47	.3918

Table A-1 *(Continued)*

z	(A) Percentile rank (percent of area below z)	(B) Percent of area from mean to z	(C) Percent of area beyond z	(D) Height of ordinate at z
0.20	57.93	07.93	42.07	.3910
0.21	58.32	08.32	41.68	.3902
0.22	58.71	08.71	41.29	.3894
0.23	59.10	09.10	40.90	.3885
0.24	59.48	09.48	40.52	.3876
0.25	59.87	09.87	40.13	.3867
0.26	60.26	10.26	39.74	.3857
0.27	60.64	10.64	39.36	.3847
0.28	61.03	11.03	38.97	.3836
0.29	61.41	11.41	38.59	.3825
0.30	61.79	11.79	38.21	.3814
0.31	62.17	12.17	37.83	.3802
0.32	62.55	12.55	37.45	.3790
0.33	62.93	12.93	37.07	.3778
0.34	63.31	13.31	36.69	.3765
0.35	63.68	13.68	36.32	.3752
0.36	64.06	14.06	35.94	.3739
0.37	64.43	14.43	35.57	.3725
0.38	64.80	14.80	35.20	.3712
0.39	65.17	15.17	34.83	.3697
0.40	65.54	15.54	34.46	.3683
0.41	65.91	15.91	34.09	.3668
0.42	66.28	16.28	33.72	.3653
0.43	66.64	16.64	33.36	.3637
0.44	67.00	17.00	33.00	.3621
0.45	67.36	17.36	32.64	.3605
0.46	67.72	17.72	32.28	.3589
0.47	68.08	18.08	31.92	.3572
0.48	68.44	18.44	31.56	.3555
0.49	68.79	18.79	31.21	.3538
0.50	69.15	19.15	30.85	.3521
0.51	69.50	19.50	30.50	.3503
0.52	69.85	19.85	30.15	.3485
0.53	70.19	20.19	29.81	.3467
0.54	70.54	20.54	29.46	.3448
0.55	70.88	20.88	29.12	.3429
0.56	71.23	21.23	28.77	.3410
0.57	71.57	21.57	28.43	.3391
0.58	71.90	21.90	28.10	.3372
0.59	72.24	22.24	27.76	.3352
0.60	72.57	22.57	27.43	.3332
0.61	72.91	22.91	27.09	.3312
0.62	73.24	23.24	26.76	.3292
0.63	73.57	23.57	26.43	.3271
0.64	73.89	23.89	26.11	.3251
0.65	74.22	24.22	25.78	.3230
0.66	74.54	24.54	25.46	.3209
0.67	74.86	24.86	25.14	.3187
0.68	75.17	25.17	24.83	.3166
0.69	75.49	25.49	24.51	.3144

Table A-1 *(Continued)*

z	(A) Percentile rank (percent of area below z)	(B) Percent of area from mean to z	(C) Percent of area beyond z	(D) Height of ordinate at z
0.70	75.80	25.80	24.20	.3123
0.71	76.11	26.11	23.89	.3101
0.72	76.42	26.42	23.58	.3079
0.73	76.73	26.73	23.27	.3056
0.74	77.04	27.04	22.96	.3034
0.75	77.34	27.34	22.66	.3011
0.76	77.64	27.64	22.36	.2989
0.77	77.94	27.94	22.06	.2966
0.78	78.23	28.23	21.77	.2943
0.79	78.52	28.52	21.48	.2920
0.80	78.81	28.81	21.19	.2897
0.81	79.10	29.10	20.90	.2874
0.82	79.39	29.39	20.61	.2850
0.83	79.67	29.67	20.33	.2827
0.84	79.95	29.95	20.05	.2803
0.85	80.23	30.23	19.77	.2780
0.86	80.51	30.51	19.49	.2756
0.87	80.78	30.78	19.22	.2732
0.88	81.06	31.06	18.94	.2709
0.89	81.33	31.33	18.67	.2685
0.90	81.59	31.59	18.41	.2661
0.91	81.86	31.86	18.14	.2637
0.92	82.12	32.12	17.88	.2613
0.93	82.38	32.38	17.62	.2589
0.94	82.64	32.64	17.36	.2565
0.95	82.89	32.89	17.11	.2541
0.96	83.15	33.15	16.85	.2516
0.97	83.40	33.40	16.60	.2492
0.98	83.65	33.65	16.35	.2468
0.99	83.89	33.89	16.11	.2444
1.00	84.13	34.13	15.87	.2420
1.01	84.38	34.38	15.62	.2396
1.02	84.61	34.61	15.39	.2371
1.03	84.85	34.85	15.15	.2347
1.04	85.08	35.08	14.92	.2323
1.05	85.31	35.31	14.69	.2299
1.06	85.54	35.54	14.46	.2275
1.07	85.77	35.77	14.23	.2251
1.08	85.99	35.99	14.01	.2227
1.09	86.21	36.21	13.79	.2203
1.10	86.43	36.43	13.57	.2179
1.11	86.65	36.65	13.35	.2155
1.12	86.86	36.86	13.14	.2131
1.13	87.08	37.08	12.92	.2107
1.14	87.29	37.29	12.71	.2083
1.15	87.49	37.49	12.51	.2059
1.16	87.70	37.70	12.30	.2036
1.17	87.90	37.90	12.10	.2012
1.18	88.10	38.10	11.90	.1989
1.19	88.30	38.30	11.70	.1965

Table A-1 *(Continued)*

z	(A) Percentile rank (percent of area below z)	(B) Percent of area from mean to z	(C) Percent of area beyond z	(D) Height of ordinate at z
1.20	88.49	38.49	11.51	.1942
1.21	88.69	38.69	11.31	.1919
1.22	88.88	38.88	11.12	.1895
1.23	89.07	39.07	10.93	.1872
1.24	89.25	39.25	10.75	.1849
1.25	89.44	39.44	10.56	.1826
1.26	89.62	39.62	10.38	.1804
1.27	89.80	39.80	10.20	.1781
1.28	89.97	39.97	10.03	.1758
1.29	90.15	40.15	09.85	.1736
1.30	90.32	40.32	09.68	.1714
1.31	90.49	40.49	09.51	.1691
1.32	90.66	40.66	09.34	.1669
1.33	90.82	40.82	09.18	.1647
1.34	90.99	40.99	09.01	.1626
1.35	91.15	41.15	08.85	.1604
1.36	91.31	41.31	08.69	.1582
1.37	91.47	41.47	08.53	.1561
1.38	91.62	41.62	08.38	.1539
1.39	91.77	41.77	08.23	.1518
1.40	91.92	41.92	08.08	.1497
1.41	92.07	42.07	07.93	.1476
1.42	92.22	42.22	07.78	.1456
1.43	92.36	42.36	07.64	.1435
1.44	92.51	42.51	07.49	.1415
1.45	92.65	42.65	07.35	.1394
1.46	92.79	42.79	07.21	.1374
1.47	92.92	42.92	07.08	.1354
1.48	93.06	43.06	06.94	.1334
1.49	93.19	43.19	06.81	.1315
1.50	93.32	43.32	06.68	.1295
1.51	93.45	43.45	06.55	.1276
1.52	93.57	43.57	06.43	.1257
1.53	93.70	43.70	06.30	.1238
1.54	93.82	43.82	06.18	.1219
1.55	93.94	43.94	06.06	.1200
1.56	94.06	44.06	05.94	.1182
1.57	94.18	44.18	05.82	.1163
1.58	94.29	44.29	05.71	.1145
1.59	94.41	44.41	05.59	.1127
1.60	94.52	44.52	05.48	.1109
1.61	94.63	44.63	05.37	.1092
1.62	94.74	44.74	05.26	.1074
1.63	94.84	44.84	05.16	.1057
1.64	94.95	44.95	05.05	.1040
1.65	95.05	45.05	04.95	.1023
1.66	95.15	45.15	04.85	.1006
1.67	95.25	45.25	04.75	.0989
1.68	95.35	45.35	04.65	.0973
1.69	95.45	45.45	04.55	.0957

Table A-1   *(Continued)*

z	(A) Percentile rank (percent of area below z)	(B) Percent of area from mean to z	(C) Percent of area beyond z	(D) Height of ordinate at z
1.70	95.54	45.54	04.46	.0940
1.71	95.64	45.64	04.36	.0925
1.72	95.73	45.73	04.27	.0909
1.73	95.82	45.82	04.18	.0893
1.74	95.91	45.91	04.09	.0878
1.75	95.99	45.99	04.01	.0863
1.76	96.08	46.08	03.92	.0848
1.77	96.16	46.16	03.84	.0833
1.78	96.25	46.25	03.75	.0818
1.79	96.33	46.33	03.67	.0804
1.80	96.41	46.41	03.59	.0790
1.81	96.49	46.49	03.51	.0775
1.82	96.56	46.56	03.44	.0761
1.83	96.64	46.64	03.36	.0748
1.84	96.71	46.71	03.29	.0734
1.85	96.78	46.78	03.22	.0721
1.86	96.86	46.86	03.14	.0707
1.87	96.93	46.93	03.07	.0694
1.88	96.99	46.99	03.01	.0681
1.89	97.06	47.06	02.94	.0669
1.90	97.13	47.13	02.87	.0656
1.91	97.19	47.19	02.81	.0644
1.92	97.26	47.26	02.74	.0632
1.93	97.32	47.32	02.68	.0620
1.94	97.38	47.38	02.62	.0608
1.95	97.44	47.44	02.56	.0596
1.96	97.50	47.50	02.50	.0584
1.97	97.56	47.56	02.44	.0573
1.98	97.61	47.61	02.39	.0562
1.99	97.67	47.67	02.33	.0551
2.00	97.72	47.72	02.28	.0540
2.01	97.78	47.78	02.22	.0529
2.02	97.83	47.83	02.17	.0519
2.03	97.88	47.88	02.12	.0508
2.04	97.93	47.93	02.07	.0498
2.05	97.98	47.98	02.02	.0488
2.06	98.03	48.03	01.97	.0478
2.07	98.08	48.08	01.92	.0468
2.08	98.12	48.12	01.88	.0459
2.09	98.17	48.17	01.83	.0449
2.10	98.21	48.21	01.79	.0440
2.11	98.26	48.26	01.74	.0431
2.12	98.30	48.30	01.70	.0422
2.13	98.34	48.34	01.66	.0413
2.14	98.38	48.38	01.62	.0404
2.15	98.42	48.42	01.58	.0396
2.16	98.46	48.46	01.54	.0387
2.17	98.50	48.50	01.50	.0379
2.18	98.54	48.54	01.46	.0371
2.19	98.57	48.57	01.43	.0363

Table A-1 *(Continued)*

z	(A) Percentile rank (percent of area below z)	(B) Percent of area from mean to z	(C) Percent of area beyond z	(D) Height of ordinate at z
2.20	98.61	48.61	01.39	.0355
2.21	98.64	48.64	01.36	.0347
2.22	98.68	48.68	01.32	.0339
2.23	98.71	48.71	01.29	.0332
2.24	98.75	48.75	01.25	.0325
2.25	98.78	48.78	01.22	.0317
2.26	98.81	48.81	01.19	.0310
2.27	98.84	48.84	01.16	.0303
2.28	98.87	48.87	01.13	.0297
2.29	98.90	48.90	01.10	.0290
2.30	98.93	48.93	01.07	.0283
2.31	98.96	48.96	01.04	.0277
2.32	98.98	48.98	01.02	.0270
2.33	99.01	49.01	00.99	.0264
2.34	99.04	49.04	00.96	.0258
2.35	99.06	49.06	00.94	.0252
2.36	99.09	49.09	00.91	.0246
2.37	99.11	49.11	00.89	.0241
2.38	99.13	49.13	00.87	.0235
2.39	99.16	49.16	00.84	.0229
2.40	99.18	49.18	00.82	.0224
2.41	99.20	49.20	00.80	.0219
2.42	99.22	49.22	00.78	.0213
2.43	99.25	49.25	00.75	.0208
2.44	99.27	49.27	00.73	.0203
2.45	99.29	49.29	00.71	.0198
2.46	99.31	49.31	00.69	.0194
2.47	99.32	49.32	00.68	.0189
2.48	99.34	49.34	00.66	.0184
2.49	99.36	49.36	00.64	.0180
2.50	99.38	49.38	00.62	.0175
2.51	99.40	49.40	00.60	.0171
2.52	99.41	49.41	00.59	.0167
2.53	99.43	49.43	00.57	.0163
2.54	99.45	49.45	00.55	.0158
2.55	99.46	49.46	00.54	.0154
2.56	99.48	49.48	00.52	.0151
2.57	99.49	49.49	00.51	.0147
2.58	99.51	49.51	00.49	.0143
2.59	99.52	49.52	00.48	.0139
2.60	99.53	49.53	00.47	.0136
2.61	99.55	49.55	00.45	.0132
2.62	99.56	49.56	00.44	.0129
2.63	99.57	49.57	00.43	.0126
2.64	99.59	49.59	00.41	.0122
2.65	99.60	49.60	00.40	.0119
2.66	99.61	49.61	00.39	.0116
2.67	99.62	49.62	00.38	.0113
2.68	99.63	49.63	00.37	.0110
2.69	99.64	49.64	00.36	.0107

Table A-1   *(Continued)*

z	(A) Percentile rank (percent of area below z)	(B) Percent of area from mean to z	(C) Percent of area beyond z	(D) Height of ordinate at z
2.70	99.65	49.65	00.35	.0104
2.71	99.66	49.66	00.34	.0101
2.72	99.67	49.67	00.33	.0099
2.73	99.68	49.68	00.32	.0096
2.74	99.69	49.69	00.31	.0093
2.75	99.70	49.70	00.30	.0091
2.76	99.71	49.71	00.29	.0088
2.77	99.72	49.72	00.28	.0086
2.78	99.73	49.73	00.27	.0084
2.79	99.74	49.74	00.26	.0081
2.80	99.74	49.74	00.26	.0079
2.81	99.75	49.75	00.25	.0077
2.82	99.76	49.76	00.24	.0075
2.83	99.77	49.77	00.23	.0073
2.84	99.77	49.77	00.23	.0071
2.85	99.78	49.78	00.22	.0069
2.86	99.79	49.79	00.21	.0067
2.87	99.79	49.79	00.21	.0065
2.88	99.80	49.80	00.20	.0063
2.89	99.81	49.81	00.19	.0061
2.90	99.81	49.81	00.19	.0060
2.91	99.82	49.82	00.18	.0058
2.92	99.82	49.82	00.18	.0056
2.93	99.83	49.83	00.17	.0055
2.94	99.84	49.84	00.16	.0053
2.95	99.84	49.84	00.16	.0051
2.96	99.85	49.85	00.15	.0051
2.97	99.85	49.85	00.15	.0048
2.98	99.86	49.86	00.14	.0047
2.99	99.86	49.86	00.14	.0046
3.00	99.87	49.87	00.13	.0044
3.01	99.87	49.87	00.13	.0043
3.02	99.87	49.87	00.13	.0042
3.03	99.88	49.88	00.12	.0040
3.04	99.88	49.88	00.12	.0039
3.05	99.89	49.89	00.11	.0038
3.06	99.89	49.89	00.11	.0037
3.07	99.89	49.89	00.11	.0036
3.08	99.90	49.90	00.10	.0035
3.09	99.90	49.90	00.10	.0034
3.10	99.90	49.90	00.10	.0033
3.11	99.91	49.91	00.09	.0032
3.12	99.91	49.91	00.09	.0031
3.13	99.91	49.91	00.09	.0030
3.14	99.92	49.92	00.08	.0029
3.15	99.92	49.92	00.08	.0028
3.16	99.92	49.92	00.08	.0027
3.17	99.92	49.92	00.08	.0026
3.18	99.93	49.93	00.07	.0025
3.19	99.93	49.93	00.07	.0025

Table A-1 *(Continued)*

z	(A) Percentile rank (percent of area below z)	(B) Percent of area from mean to z	(C) Percent of area beyond z	(D) Height of ordinate at z
3.20	99.93	49.93	00.07	.0024
3.21	99.93	49.93	00.07	.0023
3.22	99.94	49.94	00.06	.0022
3.23	99.94	49.94	00.06	.0022
3.24	99.94	49.94	00.06	.0021
3.30	99.95	49.95	00.05	.0017
3.40	99.97	49.97	00.03	.0012
3.50	99.98	49.98	00.02	.0009
3.60	99.98	49.98	00.02	.0006
3.70	99.99	49.99	00.01	.0004

Table A-2   Functions of r

r	$\sqrt{r}$	r^2	$\sqrt{r-r^2}$	$\sqrt{1-r}$	$1-r^2$	$\sqrt{1-r^2}$	$100(1-k)$	r
						k	% Eff.	
1.00	1.0000	1.0000	0.0000	0.0000	0.0000	0.0000	100.00	1.00
.99	.9950	.9801	.0995	.1000	.0199	.1411	85.89	.99
.98	.9899	.9604	.1400	.1414	.0396	.1990	80.10	.98
.97	.9849	.9409	.1706	.1732	.0591	.2431	75.69	.97
.96	.9798	.9216	.1960	.2000	.0784	.2800	72.00	.96
.95	.9747	.9025	.2179	.2236	.0975	.3122	68.78	.95
.94	.9695	.8836	.2375	.2449	.1164	.3412	65.88	.94
.93	.9644	.8649	.2551	.2646	.1351	.3676	63.24	.93
.92	.9592	.8464	.2713	.2828	.1536	.3919	60.81	.92
.91	.9539	.8281	.2862	.3000	.1719	.4146	58.54	.91
.90	.9487	.8100	.3000	.3162	.1900	.4359	56.41	.90
.89	.9434	.7921	.3129	.3317	.2079	.4560	54.40	.89
.88	.9381	.7744	.3250	.3464	.2256	.4750	52.50	.88
.87	.9327	.7569	.3363	.3606	.2431	.4931	50.69	.87
.86	.9274	.7396	.3470	.3742	.2604	.5103	48.97	.86
.85	.9220	.7225	.3571	.3873	.2775	.5268	47.32	.85
.84	.9165	.7056	.3666	.4000	.2944	.5426	45.74	.84
.83	.9110	.6889	.3756	.4123	.3111	.5578	44.22	.83
.82	.9055	.6724	.3842	.4243	.3276	.5724	42.76	.82
.81	.9000	.6561	.3923	.4359	.3439	.5864	41.36	.81
.80	.8944	.6400	.4000	.4472	.3600	.6000	40.00	.80
.79	.8888	.6241	.4073	.4583	.3759	.6131	38.69	.79
.78	.8832	.6084	.4142	.4690	.3916	.6258	37.42	.78
.77	.8775	.5929	.4208	.4796	.4071	.6380	36.20	.77
.76	.8718	.5776	.4271	.4899	.4224	.6499	35.01	.76
.75	.8660	.5625	.4300	.5000	.4375	.6614	33.86	.75
.74	.8602	.5476	.4386	.5099	.4524	.6726	32.74	.74
.73	.8544	.5329	.4440	.5196	.4671	.6834	31.66	.73
.72	.8485	.5184	.4490	.5292	.4816	.6940	30.60	.72
.71	.8426	.5041	.4538	.5385	.4959	.7042	29.58	.71
.70	.8367	.4900	.4583	.5477	.5100	.7141	28.59	.70
.69	.8307	.4761	.4625	.5568	.5239	.7238	27.62	.69
.68	.8246	.4624	.4665	.5657	.5376	.7332	26.68	.68
.67	.8185	.4489	.4702	.5745	.5511	.7424	25.76	.67
.66	.8124	.4356	.4737	.5831	.5644	.7513	24.87	.66
.65	.8062	.4225	.4770	.5916	.5775	.7599	24.01	.65
.64	.8000	.4096	.4800	.6000	.5904	.7684	23.16	.64
.63	.7937	.3969	.4828	.6083	.6031	.7766	22.34	.63
.62	.7874	.3844	.4854	.6164	.6156	.7846	21.54	.62
.61	.7810	.3721	.4877	.6245	.6279	.7924	20.76	.61
.60	.7746	.3600	.4899	.6325	.6400	.8000	20.00	.60
.59	.7681	.3481	.4918	.6403	.6519	.8074	19.26	.59
.58	.7616	.3364	.4936	.6481	.6636	.8146	18.54	.58
.57	.7550	.3249	.4951	.6557	.6751	.8216	17.84	.57
.56	.7483	.3136	.4964	.6633	.6864	.8285	17.15	.56
.55	.7416	.3025	.4975	.6708	.6975	.8352	16.48	.55
.54	.7348	.2916	.4984	.6782	.7084	.8417	15.83	.54
.53	.7280	.2809	.4991	.6856	.7191	.8480	15.20	.53
.52	.7211	.2704	.4996	.6928	.7296	.8542	14.58	.52
.51	.7141	.2601	.4999	.7000	.7399	.8602	13.98	.51

Table A-2 *(Continued)*

r	$\sqrt{r}$ ·	r^2	$\sqrt{r-r^2}$	$\sqrt{1-r}$	$1-r^2$	$\sqrt{1-r^2}$	$100(1-k)$	r
						k	% Eff.	
.50	.7071	.2500	.5000	.7071	.7500	.8660	13.40	.50
.49	.7000	.2401	.4999	.7141	.7599	.8717	12.83	.49
.48	.6928	.2304	.4996	.7211	.7696	.8773	12.27	.48
.47	.6856	.2209	.4991	.7280	.7791	.8827	11.73	.47
.46	.6782	.2116	.4984	.7348	.7884	.8879	11.21	.46
.45	.6708	.2025	.4975	.7416	.7975	.8930	10.70	.45
.44	.6633	.1936	.4964	.7483	.8064	.8980	10.20	.44
.43	.6557	.1849	.4951	.7550	.8151	.9028	9.72	.43
.42	.6481	.1764	.4936	.7616	.8236	.9075	9.25	.42
.41	.6403	.1681	.4918	.7681	.8319	.9121	8.79	.41
.40	.6325	.1600	.4899	.7746	.8400	.9165	8.35	.40
.39	.6245	.1521	.4877	.7810	.8479	.9208	7.92	.39
.38	.6164	.1444	.4854	.7874	.8556	.9250	7.50	.38
.37	.6083	.1369	.4828	.7937	.8631	.9290	7.10	.37
.36	.6000	.1296	.4800	.8000	.8704	.9330	6.70	.36
.35	.5916	.1225	.4770	.8062	.8775	.9367	6.33	.35
.34	.5831	.1156	.4737	.8124	.8844	.9404	5.96	.34
.33	.5745	.1089	.4702	.8185	.8911	.9440	5.60	.33
.32	.5657	.1024	.4665	.8246	.8976	.9474	5.25	.32
.31	.5568	.0961	.4625	.8307	.9039	.9507	4.93	.31
.30	.5477	.0900	.4583	.8367	.9100	.9539	4.61	.30
.29	.5385	.0841	.4538	.8426	.9159	.9570	4.30	.29
.28	.5292	.0784	.4490	.8485	.9216	.9600	4.00	.28
.27	.5196	.0729	.4440	.8544	.9271	.9629	3.71	.27
.26	.5099	.0676	.4386	.8602	.9324	.9656	3.44	.26
.25	.5000	.0625	.4330	.8660	.9375	.9682	3.18	.25
.24	.4899	.0576	.4271	.8718	.9424	.9708	2.92	.24
.23	.4796	.0529	.4208	.8775	.9471	.9732	2.68	.23
.22	.4690	.0484	.4142	.8832	.9516	.9755	2.45	.22
.21	.4583	.0441	.4073	.8888	.9559	.9777	2.23	.21
.20	.4472	.0400	.4000	.8944	.9600	.9798	2.02	.20
.19	.4359	.0361	.3923	.9000	.9639	.9818	1.82	.19
.18	.4243	.0324	.3842	.9055	.9676	.9837	1.63	.18
.17	.4123	.0289	.3756	.9110	.9711	.9854	1.46	.17
.16	.4000	.0256	.3666	.9165	.9744	.9871	1.29	.16
.15	.3873	.0225	.3571	.9220	.9775	.9887	1.13	.15
.14	.3742	.0196	.3470	.9274	.9804	.9902	.98	.14
.13	.3606	.0169	.3363	.9327	.9831	.9915	.85	.13
.12	.3464	.0144	.3250	.9381	.9856	.9928	.72	.12
.11	.3317	.0121	.3129	.9434	.9879	.9939	.61	.11
.10	.3162	.0100	.3000	.9487	.9900	.9950	.50	.10
.09	.3000	.0081	.2862	.9539	.9919	.9959	.41	.09
.08	.2828	.0064	.2713	.9592	.9936	.9968	.32	.08
.07	.2646	.0049	.2551	.9644	.9951	.9975	.25	.07
.06	.2449	.0036	.2375	.9695	.9964	.9982	.18	.06
.05	.2236	.0025	.2179	.9747	.9975	.9987	.13	.05
.04	.2000	.0016	.1960	.9798	.9984	.9992	.08	.04
.03	.1732	.0009	.1706	.9849	.9991	.9995	.05	.03
.02	.1414	.0004	.1400	.9899	.9996	.9998	.02	.02
.01	.1000	.0001	.0995	.9950	.9999	.9999	.01	.01
.00	.0000	.0000	.0000	1.0000	1.0000	1.0000	.00	.00

Table A-3   Squares, square roots, and reciprocals of numbers from 1
to 1000

N	N²	√N	1/N	N	N²	√N	1/N	N	N²	√N	1/N
1	1	1.0000	1.000000	61	3721	7.8102	.016393	121	14641	11.0000	.00826446
2	4	1.4142	.500000	62	3844	7.8740	.016129	122	14884	11.0454	.00819672
3	9	1.7321	.333333	63	3969	7.9373	.015873	123	15129	11.0905	.00813008
4	16	2.0000	.250000	64	4096	8.0000	.015625	124	15376	11.1355	.00800452
5	25	2.2361	.200000	65	4225	8.0623	.015385	125	15625	11.1803	.00800000
6	36	2.4495	.166667	66	4356	8.1240	.015152	126	15876	11.2250	.00793651
7	49	2.6458	.142857	67	4489	8.1854	.014925	127	16129	11.2694	.00787402
8	64	2.8284	.125000	68	4624	8.2462	.014706	128	16384	11.3137	.00781250
9	81	3.0000	.111111	69	4761	8.3066	.014493	129	16641	11.3578	.00775194
10	100	3.1623	.100000	70	4900	8.3666	.014286	130	16900	11.4018	.00769231
11	121	3.3166	.090909	71	5041	8.4261	.014085	131	17161	11.4455	.00763359
12	144	3.4641	.083333	72	5184	8.4853	.013889	132	17424	11.4891	.00757576
13	169	3.6056	.076923	73	5329	8.5440	.013699	133	17689	11.5326	.00751880
14	196	3.7417	.071429	74	5476	8.6023	.013514	134	17956	11.5758	.00746269
15	225	3.8730	.066667	75	5625	8.6603	.013333	135	18225	11.6190	.00740741
16	256	4.0000	.062500	76	5776	8.7178	.013158	136	18496	11.6619	.00735294
17	289	4.1231	.058824	77	5929	8.7750	.012987	137	18769	11.7047	.00729927
18	324	4.2426	.055556	78	6084	8.8318	.012821	138	19044	11.7473	.00724638
19	361	4.3589	.052632	79	6241	8.8882	.012658	139	19321	11.7898	.00719424
20	400	4.4721	.050000	80	6400	8.9443	.012500	140	19600	11.8322	.00714286
21	441	4.5826	.047619	81	6561	9.0000	.012346	141	19881	11.8743	.00709220
22	484	4.6904	.045455	82	6724	9.0554	.012195	142	20164	11.9164	.00704225
23	529	4.7958	.043478	83	6889	9.1104	.012048	143	20449	11.9583	.00699301
24	576	4.8990	.041667	84	7056	9.1652	.011905	144	20736	12.0000	.00694444
25	625	5.0000	.040000	85	7225	9.2195	.011765	145	21025	12.0416	.00689655
26	676	5.0990	.038462	86	7396	9.2736	.011628	146	21316	12.0830	.00684932
27	729	5.1962	.037037	87	7569	9.3274	.011494	147	21609	12.1244	.00680272
28	784	5.2915	.035714	88	7744	9.3808	.011364	148	21904	12.1655	.00675676
29	841	5.3852	.034483	89	7921	9.4340	.011236	149	22201	12.2066	.00671141
30	900	5.4772	.033333	90	8100	9.4868	.011111	150	22500	12.2474	.00666667
31	961	5.5678	.032258	91	8281	9.5394	.010989	151	22801	12.2882	.00662252
32	1024	5.6569	.031250	92	8464	9.5917	.010870	152	23104	12.3288	.00657895
33	1089	5.7446	.030303	93	8649	9.6437	.010753	153	23409	12.3693	.00653595
34	1156	5.8310	.029412	94	8836	9.6954	.010638	154	23716	12.4097	.00649351
35	1225	5.9161	.028571	95	9025	9.7468	.010526	155	24025	12.4499	.00645161
36	1296	6.0000	.027778	96	9216	9.7980	.010417	156	24336	12.4900	.00641026
37	1369	6.0828	.027027	97	9409	9.8489	.010309	157	24649	12.5300	.00636943
38	1444	6.1644	.026316	98	9604	9.8995	.010204	158	24964	12.5698	.00632911
39	1521	6.2450	.025641	99	9801	9.9499	.010101	159	25281	12.6095	.00628931
40	1600	6.3246	.025000	100	10000	10.0000	.010000	160	25600	12.6491	.00625000
41	1681	6.4031	.024390	101	10201	10.0499	.00990099	161	25921	12.6886	.00621118
42	1764	6.4807	.023810	102	10404	10.0995	.00980392	162	26244	12.7279	.00617284
43	1849	6.5574	.023256	103	10609	10.1489	.00970874	163	26569	12.7671	.00613497
44	1936	6.6332	.022727	104	10816	10.1980	.00961538	164	26896	12.8062	.00609756
45	2025	6.7082	.022222	105	11025	10.2470	.00952381	165	27225	12.8452	.00606061
46	2116	6.7823	.021739	106	11236	10.2956	.00943396	166	27556	12.8841	.00602410
47	2209	6.8557	.021277	107	11449	10.3441	.00934579	167	27889	12.9228	.00598802
48	2304	6.9282	.020833	108	11664	10.3923	.00925926	168	28224	12.9615	.00595238
49	2401	7.0000	.020408	109	11881	10.4403	.00917431	169	28561	13.0000	.00591716
50	2500	7.0711	.020000	110	12100	10.4881	.00909091	170	28900	13.0384	.00588235
51	2601	7.1414	.019608	111	12321	10.5357	.00900901	171	29241	13.0767	.00584795
52	2704	7.2111	.019231	112	12544	10.5830	.00892857	172	29584	13.1149	.00581395
53	2809	7.2801	.018868	113	12769	10.6301	.00884956	173	29929	13.1529	.00578035
54	2916	7.3485	.018519	114	12996	10.6771	.00877193	174	30276	13.1909	.00574713
55	3025	7.4162	.018182	115	13225	10.7238	.00869565	175	30625	13.2288	.00571429
56	3136	7.4833	.017857	116	13456	10.7703	.00862069	176	30976	13.2665	.00568182
57	3249	7.5498	.017544	117	13689	10.8167	.00854701	177	31329	13.3041	.00564972
58	3364	7.6158	.017241	118	13924	10.8628	.00847458	178	31684	13.3417	.00561798
59	3481	7.6811	.016949	119	14161	10.9087	.00840336	179	32041	13.3791	.00558659
60	3600	7.7460	.016667	120	14400	10.9545	.00833333	180	32400	13.4164	.00555556

Table A-3 *(Continued)*

N	N²	√N	1/N	N	N²	√N	1/N	N	N²	√N	1/N
181	32761	13.4536	.00552486	241	58081	15.5242	.00414938	301	90601	17.3494	.00332226
182	33124	13.4907	.00549451	242	58564	15.5563	.00413223	302	91204	17.3781	.00331126
183	33489	13.5277	.00546448	243	59049	15.5885	.00411523	303	91809	17.4069	.00330033
184	33856	13.5647	.00543478	244	59536	15.6205	.00409836	304	92416	17.4356	.00328047
185	34225	13.6015	.00540541	245	60025	15.6525	.00408163	305	93025	17.4642	.00328947
186	34596	13.6382	.00537634	246	60516	15.6844	.00406504	306	93636	17.4929	.00326797
187	34969	13.6748	.00534759	247	61009	15.7162	.00404858	307	94249	17.5214	.00325733
188	35344	13.7113	.00531915	248	61504	15.7480	.00403226	308	94864	17.5499	.00321675
189	35721	13.7477	.00529101	249	62001	15.7797	.00401606	309	95481	17.5784	.00323625
190	36100	13.7840	.00526316	250	62500	15.8114	.00400000	310	96100	17.6068	.00322581
191	36481	13.8203	.00523560	251	63001	15.8430	.00398406	311	96721	17.6352	.00321543
192	36864	13.8564	.00520833	252	63504	15.8745	.00396825	312	97344	17.6635	.00320513
193	37249	13.8924	.00518135	253	64009	15.9060	.00395257	313	97969	17.6918	.00319489
194	37636	13.9284	.00515464	254	64516	15.9374	.00393701	314	98596	17.7200	.00318471
195	38025	13.9642	.00512821	255	65025	15.9687	.00392157	315	99225	17.7482	.00317460
196	38416	14.0000	.00510204	256	65536	16.0000	.00390625	316	99856	17.7764	.00316456
197	38809	14.0357	.00507614	257	66049	16.0312	.00389105	317	100489	17.8045	.00315457
198	39204	14.0712	.00505051	258	66564	16.0624	.00387597	318	101124	17.8326	.00314465
199	39601	14.1067	.00502513	259	67081	16.0935	.00386100	319	101761	17.8606	.00313480
200	40000	14.1421	.00500000	260	67600	16.1245	.00384615	320	102400	17.8885	.00312500
201	40401	14.1774	.00497512	261	68121	16.1555	.00383142	321	103041	17.9165	.00311526
202	40804	14.2127	.00495050	262	68644	16.1864	.00381679	322	103684	17.9444	.00310559
203	41209	14.2478	.00492611	263	69169	16.2173	.00380228	323	104329	17.9722	.00309598
204	41616	14.2829	.00490196	264	69696	16.2481	.00378788	324	104976	18.0000	.00308642
205	42025	14.3178	.00487805	265	70225	16.2788	.00377358	325	105625	18.0278	.00307692
206	42436	14.3527	.00485437	266	70756	16.3095	.00375940	326	106276	18.0555	.00306748
207	42849	14.3875	.00483092	267	71289	16.3401	.00374532	327	106929	18.0831	.00305810
208	43264	14.4222	.00480769	268	71824	16.3707	.00373134	328	107584	18.1108	.00304878
209	43681	14.4568	.00478469	269	72361	16.4012	.00371747	329	108241	18.1384	.00303951
210	44100	14.4914	.00476190	270	72900	16.4317	.00370370	330	108900	18.1659	.00303030
211	44521	14.5258	.00473934	271	73441	16.4621	.00369004	331	109561	18.1934	.00302115
212	44944	14.5602	.00471698	272	73984	16.4924	.00367647	332	110224	18.2209	.00301205
213	45369	14.5945	.00469484	273	74529	16.5227	.00366300	333	110889	18.2483	.00300300
214	45796	14.6287	.00467290	274	75076	16.5529	.00364964	334	111556	18.2757	.00299401
215	46225	14.6629	.00465116	275	75625	16.5831	.00363636	335	112225	18.3030	.00298507
216	46656	14.6969	.00462963	276	76176	16.6132	.00362319	336	112896	18.3303	.00297619
217	47089	14.7309	.00460829	277	76729	16.6433	.00361011	337	113569	18.3576	.00296736
218	47524	14.7648	.00458716	278	77284	16.6733	.00359712	338	114244	18.3848	.00295858
219	47961	14.7986	.00456621	279	77841	16.7033	.00358423	339	114921	18.4120	.00294985
220	48400	14.8324	.00454545	280	78400	16.7332	.00357143	340	115600	18.4391	.00294118
221	48841	14.8661	.00452489	281	78961	16.7631	.00355872	341	116281	18.4662	.00293255
222	49284	14.8997	.00450450	282	79524	16.7929	.00354610	342	116964	18.4932	.00292398
223	49729	14.9332	.00448430	283	80089	16.8226	.00353357	343	117649	18.5203	.00291545
224	50176	14.9666	.00446429	284	80656	16.8523	.00352113	344	118336	18.5472	.00290698
225	50625	15.0000	.00444444	285	81225	16.8819	.00350877	345	119025	18.5742	.00289855
226	51076	15.0333	.00442478	286	81796	16.9115	.00349650	346	119716	18.6011	.00289017
227	51529	15.0665	.00440529	287	82369	16.9411	.00348432	347	120409	18.6279	.00288184
228	51984	15.0997	.00438596	288	82944	16.9706	.00347222	348	121104	18.6548	.00287356
229	52441	15.1327	.00436681	289	83521	17.0000	.00346021	349	121801	18.6815	.00286533
230	52900	15.1658	.00434783	290	84100	17.0294	.00344828	350	122500	18.7083	.00285714
231	53361	15.1987	.00432900	291	84681	17.0587	.00343643	351	123201	18.7350	.00284900
232	53824	15.2315	.00431034	292	85264	17.0880	.00342466	352	123904	18.7617	.00284091
233	54289	15.2643	.00429185	293	85849	17.1172	.00341297	353	124609	18.7883	.00283286
234	54756	15.2971	.00427350	294	86436	17.1464	.00340136	354	125316	18.8149	.00282486
235	55225	15.3297	.00425532	295	87025	17.1756	.00338983	355	126025	18.8414	.00281690
236	55696	15.3623	.00423729	296	87616	17.2047	.00337838	356	126736	18.8680	.00280899
237	56169	15.3948	.00421941	297	88209	17.2337	.00336700	357	127449	18.8944	.00280112
238	56644	15.4272	.00420168	298	88804	17.2627	.00335570	358	128164	18.9209	.00279330
239	57121	15.4596	.00418410	299	89401	17.2916	.00334448	359	128881	18.9473	.00278552
240	57600	15.4919	.00416667	300	90000	17.3205	.00333333	360	129600	18.9737	.00277778

Table A-3 *(Continued)*

N	N²	√N	1/N	N	N²	√N	1/N	N	N²	√N	1/N
361	130321	19.0000	.00277008	421	177241	20.5183	.00237530	481	231361	21.9317	.00207900
362	131044	19.0263	.00276243	422	178084	20.5426	.00236967	482	232324	21.9545	.00207469
363	131769	19.0526	.00275482	423	178929	20.5670	.00236407	483	233289	21.9773	.00207039
364	132496	19.0788	.00274725	424	179776	20.5913	.00235849	484	234256	22.0000	.00206612
365	133225	19.1050	.00273973	425	180625	20.6155	.00235294	485	235225	22.0227	.00206186
366	133956	19.1311	.00273224	426	181476	20.6398	.00234742	486	236196	22.0454	.00205761
367	134689	19.1572	.00272480	427	182329	20.6640	.00234192	487	237169	22.0681	.00205339
368	135424	19.1833	.00271739	428	183184	20.6882	.00233645	488	238144	22.0907	.00204918
369	136161	19.2094	.00271003	429	184041	20.7123	.00233100	489	239121	22.1133	.00204499
370	136900	19.2354	.00270270	430	184900	20.7364	.00232558	490	240100	22.1359	.00204082
371	137641	19.2614	.00269542	431	185761	20.7605	.00232019	491	241081	22.1585	.00203666
372	138384	19.2873	.00268817	432	186624	20.7846	.00231481	492	242064	22.1811	.00203252
373	139129	19.3132	.00268097	433	187489	20.8087	.00230947	493	243049	22.2036	.00202840
374	139876	19.3391	.00267380	434	188356	20.8327	.00230415	494	244036	22.2261	.00202429
375	140625	19.3649	.00266667	435	189225	20.8567	.00229885	495	245025	22.2486	.00202020
376	141376	19.3907	.00265957	436	190096	20.8806	.00229358	496	246016	22.2711	.00201613
377	142129	19.4165	.00265252	437	190969	20.9045	.00228833	497	247009	22.2935	.00201207
378	142884	19.4422	.00264550	438	191844	20.9284	.00228311	498	248004	22.3159	.00200803
379	143641	19.4679	.00263852	439	192721	20.9523	.00227790	499	249001	22.3383	.00200401
380	144400	19.4936	.00263158	440	193600	20.9762	.00227273	500	250000	22.3607	.00200000
381	145161	19.5192	.00262467	441	194481	21.0000	.00226757	501	251001	22.3830	.00199601
382	145924	19.5448	.00261780	442	195364	21.0238	.00226244	502	252004	22.4054	.00199203
383	146689	19.5704	.00261097	443	196249	21.0476	.00225734	503	253009	22.4277	.00198807
384	147456	19.5959	.00260417	444	197136	21.0713	.00225225	504	254016	22.4499	.00198413
385	148225	19.6214	.00259740	445	198025	21.0950	.00224719	505	255025	22.4722	.00198020
386	148996	19.6469	.00259067	446	198916	21.1187	.00224215	506	256036	22.4944	.00197628
387	149769	19.6723	.00258398	447	199809	21.1424	.00223714	507	257049	22.5167	.00197239
388	150544	19.6977	.00257732	448	200704	21.1660	.00223214	508	258064	22.5389	.00196850
389	151321	19.7231	.00257069	449	201601	21.1896	.00222717	509	259081	22.5610	.00196464
390	152100	19.7484	.00256410	450	202500	21.2132	.00222222	510	260100	22.5832	.00196078
391	152881	19.7737	.00255754	451	203401	21.2368	.00221729	511	261121	22.6053	.00195695
392	153664	19.7990	.00255102	452	204304	21.2603	.00221239	512	262144	22.6274	.00195312
393	154449	19.8242	.00254453	453	205209	21.2838	.00220751	513	263169	22.6495	.00194932
394	155236	19.8494	.00253807	454	206116	21.3073	.00220264	514	264196	22.6716	.00194553
395	156025	19.8746	.00253165	455	207025	21.3307	.00219870	515	265225	22.6936	.00194175
396	156816	19.8997	.00252525	456	207936	21.3542	.00219298	516	266256	22.7156	.00193798
397	157609	19.9249	.00251889	457	208849	21.3776	.00218818	517	267289	22.7376	.00193424
398	158404	19.9499	.00251256	458	209764	21.4009	.00218341	518	268324	22.7596	.00193050
399	159201	19.9750	.00250627	459	210681	21.4243	.00217865	519	269361	22.7816	.00192678
400	160000	20.0000	.00250000	460	211600	21.4476	.00217391	520	270400	22.8035	.00192308
401	160801	20.0250	.00249377	461	212521	21.4709	.00216920	521	271441	22.8254	.00191939
402	161604	20.0499	.00248756	462	213444	21.4942	.00216450	522	272484	22.8473	.00191571
403	162409	20.0749	.00248139	463	214369	21.5174	.00215983	523	273529	22.8692	.00191205
404	163216	20.0998	.00247525	464	215296	21.5407	.00215517	524	274576	22.8910	.00190840
405	164025	20.1246	.00246914	465	216225	21.5639	.00215054	525	275625	22.9129	.00190476
406	164836	20.1494	.00246305	466	217156	21.5870	.00214592	526	276676	22.9347	.00190114
407	165649	20.1742	.00245700	467	218089	21.6102	.00214133	527	277729	22.9565	.00189753
408	166464	20.1990	.00245098	468	219024	21.6333	.00213675	528	278784	22.9783	.00189394
409	167281	20.2237	.00244499	469	219961	21.6564	.00213220	529	279841	23.0000	.00189036
410	168100	20.2485	.00243902	470	220900	21.6795	.00212766	530	280900	23.0217	.00188679
411	168921	20.2731	.00243309	471	221841	21.7025	.00212314	531	281961	23.0434	.00188324
412	169744	20.2978	.00242718	472	222784	21.7256	.00211864	532	283024	23.0651	.00187970
413	170569	20.3224	.00242131	473	223729	21.7486	.00211416	533	284089	23.0868	.00187617
414	171396	20.3470	.00241546	474	224676	21.7715	.00210970	534	285156	23.1084	.00187266
415	172225	20.3715	.00240964	475	225625	21.7945	.00210526	535	286225	23.1301	.00186916
416	173056	20.3961	.00240385	476	226576	21.8174	.00210084	536	287296	23.1517	.00186567
417	173889	20.4206	.00239808	477	227529	21.8403	.00209644	537	288369	23.1733	.00186220
418	174724	20.4450	.00239234	478	228484	21.8632	.00209205	538	289444	23.1948	.00185874
419	175561	20.4695	.00238663	479	229441	21.8861	.00208768	539	290521	23.2164	.00185529
420	176400	20.4939	.00238095	480	230400	21.9089	.00208333	540	291600	23.2379	.00185185

Table A-3 *(Continued)*

N	N²	√N	1/N	N	N²	√N	1/N	N	N²	√N	1/N
541	292681	23.2594	.00184843	601	361201	24.5153	.00166389	661	436921	25.7099	.00151286
542	293764	23.2809	.00184502	602	302404	24.5357	.00166113	662	438244	25.7294	.00151057
543	294849	23.3024	.00184162	603	363609	24.5561	.00165837	663	439569	25.7488	.00150830
544	295936	23.3238	.00183824	604	364816	24.5764	.00165563	664	440896	25.7682	.00150602
545	297025	23.3452	.00183486	605	366025	24.5967	.00165289	665	442225	25.7876	.00150376
566	298116	23.3666	.00183150	606	367236	24.6171	.00165017	666	443556	25.8070	.00150150
547	299209	23.3880	.00182815	607	368449	24.6374	.00164745	667	444889	25.8263	.00149925
548	300304	23.4094	.00182482	608	369664	24.6577	.00164474	668	446224	25.8457	.00149701
549	301401	23.4307	.00182149	609	370881	24.6779	.00164204	669	447561	25.8650	.00149477
550	302500	23.4521	.00181818	610	372100	24.6982	.00163934	670	448900	25.8844	.00149254
551	303601	23.4734	.00181488	611	373321	24.7184	.00163666	671	450241	25.9037	.00149031
552	304704	23.4947	.00181159	612	374544	24.7386	.00163399	672	451584	25.9230	.00148810
553	305809	23.5160	.00180832	613	375769	24.7588	.00163132	673	452929	25.9422	.00148588
554	306916	23.5372	.00180505	614	376996	24.7790	.00162866	674	454276	25.9615	.00148368
555	308025	23.5584	.00180180	615	378225	24.7992	.00162602	675	455625	25.9808	.00148148
556	309136	23.5797	.00179856	616	379456	24.8193	.00162338	676	456976	26.0000	.00147929
557	310249	23.6008	.00179533	617	380689	24.8395	.00162075	677	458329	26.0192	.00147710
558	311364	23.6220	.00179211	618	381924	24.8596	.00161812	678	459684	26.0384	.00147493
559	312481	23.6432	.00178891	619	383161	24.8797	.00161551	679	461041	26.0576	.00147275
560	313600	23.6643	.00178571	620	384400	24.8998	.00161290	680	462400	26.0768	.00147059
561	314721	23.6854	.00178253	621	385641	24.9199	.00161031	681	463761	26.0960	.00146843
562	315844	23.7065	.00177936	622	386884	24.9399	.00160772	682	465124	26.1151	.00146628
563	316969	23.7276	.00177620	623	388129	24.9600	.00160514	683	466489	26.1343	.00146413
564	318096	23.7487	.00177305	624	389376	24.9800	.00160256	684	467856	26.1534	.00146199
565	319225	23.7697	.00176991	625	390625	25.0000	.00160000	685	469225	26.1725	.00145985
566	320356	23.7908	.00176678	626	391876	25.0200	.00159744	686	470596	26.1916	.00145773
567	321489	23.8118	.00176367	627	393129	25.0400	.00159490	687	471969	26.2107	.00145560
568	322624	23.8328	.00176056	628	394384	25.0599	.00159236	688	473344	26.2298	.00145349
569	323761	23.8537	.00175747	629	395641	25.0799	.00158983	689	474721	26.2488	.00145138
570	324900	23.8747	.00175439	630	396900	25.0998	.00158730	690	476100	26.2679	.00144928
571	326041	23.8956	.00175131	631	398161	25.1197	.00158479	691	477481	26.2869	.00144718
572	327184	23.9165	.00164825	632	399424	25.1396	.00158228	692	478864	26.3059	.00144509
573	328329	23.9374	.00174520	633	400689	25.1595	.00157978	693	480249	26.3249	.00144300
574	329476	23.9583	.00174216	634	401956	25.1794	.00157729	694	481636	26.3439	.00144092
575	330625	23.9792	.00173913	635	403225	25.1992	.00157480	695	483025	26.3629	.00143885
576	331776	24.0000	.00173611	636	404496	25.2190	.00157233	696	484416	26.3818	.00143678
577	332929	24.0208	.00173310	637	405769	25.2389	.00156986	697	485809	26.4008	.00143472
578	334084	24.0416	.00173010	638	407044	25.2587	.00156740	698	487204	26.4197	.00143266
579	335241	24.0624	.00172712	639	408321	25.2784	.00156495	699	488601	26.4386	.00143062
580	336400	24.0832	.00172414	640	409600	25.2982	.00156250	700	490000	26.4575	.00142857
581	337561	24.1039	.00172117	641	410881	25.3180	.00156006	701	491401	26.4764	.00142653
582	338724	24.1247	.00171821	642	412164	25.3377	.00155763	702	492804	26.4953	.00142450
583	339889	24.1454	.00171527	643	413449	25.3574	.00155521	703	494209	26.5141	.00142248
584	341056	24.1661	.00171233	644	414736	25.3772	.00155280	704	495616	26.5330	.00142045
585	342225	24.1868	.00170940	645	416025	25.3969	.00155039	705	497025	26.5518	.00141844
586	343396	24.2074	.00170648	646	417316	25.4165	.00154799	706	498436	26.5707	.00141643
587	344569	24.2281	.00170358	647	418609	25.4362	.00154560	707	499849	26.5895	.00141443
588	345744	24.2487	.00170068	648	419904	25.4558	.00154321	708	501264	26.6083	.00141243
589	346921	24.2693	.00169779	649	421201	25.4755	.00154083	709	502681	26.6271	.00141044
590	348100	24.2899	.00169492	650	422500	25.4951	.00153846	710	504100	26.6458	.00140845
591	349281	24.3105	.00169205	651	423801	25.5147	.00153610	711	505521	26.6646	.00140647
592	350464	24.3311	.00168919	652	425104	25.5343	.00153374	712	506944	26.6833	.00140449
593	351649	24.3516	.00168634	653	426409	25.5539	.00153139	713	508369	26.7021	.00140252
594	352836	24.3721	.00168350	654	427716	25.5734	.00152905	714	509796	26.7208	.00140056
595	354025	24.3926	.00168067	655	429025	25.5930	.00152672	715	511225	26.7395	.00139860
596	355216	24.4131	.00167785	656	430336	25.6125	.00152439	716	512656	26.7582	.00139665
597	356409	24.4336	.00167504	657	431649	25.6320	.00152207	717	514089	26.7769	.00139470
598	357604	24.4540	.00167224	658	432964	25.6515	.00151976	718	515524	26.7955	.00139276
599	358801	24.4745	.00166945	659	434281	25.6710	.00151745	719	516961	26.8142	.00139082
600	360000	24.4949	.00166667	660	435600	25.6905	.00151515	720	518400	26.8328	.00138889

Table A-3   *(Continued)*

N	N²	√N	1/N	N	N²	√N	1/N	N	N²	√N	1/N
721	519841	26.8514	.00138696	781	609961	27.9464	.00128041	841	707281	29.0000	.00118906
722	521284	26.8701	.00138504	782	611524	27.9643	.00127877	842	708964	29.0172	.00118765
723	522729	26.8887	.00138313	783	613089	27.9821	.00127714	843	710649	29.0345	.00118624
724	524176	26.9072	.00138122	784	614656	28.0000	.00127551	844	712336	29.0517	.00118483
725	525625	26.9258	.00137931	785	616225	28.0179	.00127389	845	714025	29.0689	.00118343
726	527076	26.9444	.00137741	786	617796	28.0357	.00127226	846	715716	29.0861	.00118203
727	528529	26.9629	.00137552	787	619369	28.0535	.00127065	847	717409	29.1033	.00118064
728	529984	26.9815	.00137363	788	620944	28.0713	.00126904	848	719104	29.1204	.00117925
729	531441	27.0000	.00137174	789	622521	28.0891	.00126743	849	720801	29.1376	.00117786
730	532900	27.0185	.00136986	790	624100	28.1069	.00126582	850	722500	29.1548	.00117647
731	534361	27.0370	.00136799	791	625681	28.1247	.00126422	851	724201	29.1719	.00117509
732	535824	27.0555	.00136612	792	627264	28.1425	.00126263	852	725904	29.1890	.00117371
733	537289	27.0740	.00136426	793	628849	28.1603	.00126103	853	727609	29.2062	.00117233
734	538756	27.0924	.00136240	794	630436	28.1780	.00125945	854	729316	29.2233	.00117096
735	540225	27.1109	.00136054	795	632025	28.1957	.00125786	855	731025	29.2404	.00116959
736	541696	27.1293	.00135870	796	633616	28.2135	.00125628	856	732736	29.2575	.00116822
737	543169	27.1477	.00135685	797	635209	28.2312	.00125471	857	734449	29.2746	.00116686
738	544644	27.1662	.00135501	798	636804	28.2489	.00125313	858	736164	29.2916	.00116550
739	546121	27.1846	.00135318	799	638401	28.2666	.00125156	859	737881	29.3087	.00116414
740	547600	27.2029	.00135135	800	640000	28.2843	.00125000	860	739600	29.3258	.00116279
741	549081	27.2213	.00134953	801	641601	28.3019	.00124844	861	741321	29.3428	.00116144
742	550564	27.2397	.00134771	802	643204	28.3196	.00124688	862	743044	29.3598	.00116009
743	552049	27.2580	.00134590	803	644809	28.3373	.00124533	863	744769	29.3769	.00115875
744	553536	27.2764	.00134409	804	646416	28.3549	.00124378	864	746496	29.3939	.00115741
745	555025	27.2947	.00134228	805	648025	28.3725	.00124224	865	748225	29.4109	.00115607
746	556516	27.3130	.00134048	806	649636	28.3901	.00124069	866	749956	29.4279	.00115473
747	558009	27.3313	.00133869	807	651249	28.4077	.00123916	867	751689	29.4449	.00115340
748	559504	27.3496	.00133690	808	625864	28.4253	.00123762	868	753424	29.4618	.00115207
749	561001	27.3679	.00133511	809	654481	28.4429	.00123609	869	755161	29.4788	.00115075
750	562500	27.3861	.00133333	810	656100	28.4605	.00123457	870	756900	29.4958	.00114943
751	564001	27.4044	.00133156	811	657721	28.4781	.00123305	871	758641	29.5127	.00114811
752	565504	27.4226	.00132979	812	659344	28.4956	.00123153	872	760384	29.5296	.00114679
753	567009	27.4408	.00132802	813	660969	28.5132	.00123001	873	762129	29.5466	.00114548
754	568516	27.4591	.00132626	814	662596	28.5307	.00122850	874	763876	29.5635	.00114416
755	570025	27.4773	.00132450	815	664225	28.5482	.00122699	875	765625	29.5804	.00114286
756	571536	27.4955	.00132275	816	665856	28.5657	.00122549	876	767376	29.5973	.00114155
757	573049	27.5136	.00132100	817	667489	28.5832	.00122399	877	769129	29.6142	.00114025
758	574564	27.5318	.00131926	818	669124	28.6007	.00122249	878	770884	29.6311	.00113895
759	576081	27.5500	.00131752	819	670761	28.6182	.00122100	879	772641	29.6479	.00113766
760	577600	27.5681	.00131579	820	672400	28.6356	.00121951	880	774400	29.6648	.00113636
761	579121	27.5862	.00131406	821	674041	28.6531	.00121803	881	776161	29.6816	.00113507
762	580644	27.6043	.00131234	822	675684	28.6705	.00121655	882	777924	29.6985	.00113379
763	582169	27.6225	.00131062	823	677329	28.6880	.00121507	883	779689	29.7153	.00113250
764	583696	27.6405	.00130890	824	678976	28.7054	.00121359	884	781456	29.7321	.00113122
765	585225	27.6586	.00130719	825	680625	28.7228	.00121212	885	783225	29.7489	.00112994
766	586756	27.6767	.00130548	826	682276	28.7402	.00121065	886	784996	29.7658	.00112867
767	588289	27.6948	.00130378	827	683929	28.7576	.00120919	887	786769	29.7825	.00112740
768	589824	27.7128	.00130208	828	685584	28.7750	.00120773	888	788544	29.7993	.00112613
769	591361	27.7308	.00130039	829	687241	28.7924	.00120627	889	790321	29.8161	.00112486
770	592900	27.7489	.00129870	830	688900	28.8097	.00120482	890	792100	29.8329	.00112360
771	594441	27.7669	.00129702	831	690561	28.8271	.00120337	891	793881	29.8496	.00112233
772	595984	27.7849	.00129534	832	692224	28.8444	.00120192	892	795664	29.8664	.00112108
773	597529	27.8029	.00129366	833	693889	28.8617	.00120048	893	797449	29.8831	.00111982
774	599076	27.8209	.00129199	834	695556	28.8791	.00119904	894	799236	29.8998	.00111857
775	600625	27.8388	.00129032	835	697225	28.8964	.00119760	895	801025	29.9166	.00111732
776	602176	27.8568	.00128866	836	698896	28.9137	.00119617	896	802816	29.9333	.00111607
777	603729	27.8747	.00128700	837	700569	28.9310	.00119474	897	804609	29.9500	.00111483
778	605284	27.8927	.00128535	838	702244	28.9482	.00119332	898	806404	29.9666	.00111359
779	606841	27.9106	.00128370	839	703921	28.9655	.00119190	899	808201	29.9833	.00111235
780	608400	27.9285	.00128205	840	705600	28.9828	.00119048	900	810000	30.0000	.00111111

Table A-3 *(Continued)*

N	N²	√N	1/N	N	N²	√N	1/N	N	N²	√N	1/N
901	811801	30.0167	.00110988	936	876096	30.5941	.00106838	971	942841	31.1609	.00102987
902	813604	30.0333	.00110865	937	877969	30.6105	.00106724	972	944784	31.1769	.00102881
903	815409	30.0500	.00110742	938	879844	30.6268	.00106610	973	946729	31.1929	.00102775
904	817216	30.0666	.00110619	939	881721	30.6431	.00106496	974	948676	31.2090	.00102669
905	819025	30.0832	.00110497	940	883600	30.6594	.00106383	975	950625	31.2250	.00102564
906	820836	30.0998	.00110375	941	885481	30.6757	.00106270	976	952576	31.2410	.00102459
907	822649	30.1164	.00110254	942	887364	30.6920	.00106157	977	954529	31.2570	.00102354
908	824464	30.1330	.00110132	943	889249	30.7083	.00106045	978	956484	31.2730	.00102249
909	826281	30.1496	.00110011	944	891136	30.7246	.00105932	979	958441	31.2890	.00102145
910	828100	30.1662	.00109890	945	893025	30.7409	.00105820	980	960400	31.3050	.00102041
911	829921	30.1828	.00109769	946	894916	30.7571	.00105708	981	962361	31.3209	.00101937
912	831744	30.1993	.00109649	947	896809	30.7734	.00105597	982	964324	31.3369	.00101833
913	833569	30.2159	.00109529	948	898704	30.7896	.00105485	983	966289	31.3528	.00101729
914	835396	30.2324	.00109409	949	900601	30.8058	.00105374	984	968256	31.3688	.00101626
915	837225	30.2490	.00109290	950	902500	30.8221	.00105263	985	970225	31.3847	.00101523
916	839056	30.2655	.00109170	951	904401	30.8383	.00105152	986	972196	31.4006	.00101420
917	840889	30.2820	.00109051	952	906304	30.8545	.00105042	987	974169	31.4166	.00101317
918	842724	30.2985	.00108932	953	908209	30.8707	.00104932	988	976144	31.4325	.00101215
919	844561	30.3150	.00108814	954	910116	30.8869	.00104822	989	978121	31.4484	.00101112
920	846400	30.3315	.00108696	955	912025	30.9031	.00104712	990	980100	31.4643	.00101010
921	848241	30.3480	.00108578	956	913936	30.9192	.00104603	991	982081	31.4802	.00100908
922	850084	30.3645	.00108460	957	915849	30.9354	.00104493	992	984064	31.4960	.00100806
923	851929	30.3809	.00108342	958	917764	30.9516	.00104384	993	986049	31.5119	.00100705
924	853776	30.3974	.00108225	959	919681	30.9677	.00104275	994	988036	31.5278	.00100604
925	855625	30.4138	.00108108	960	921600	30.9839	.00104167	995	990025	31.5436	.00100503
926	857476	30.4302	.00107991	961	923521	31.0000	.00104058	996	992016	31.5595	.00100402
927	859329	30.4467	.00107875	962	925444	31.0161	.00103950	997	994009	31.5753	.00103842
928	861184	30.4631	.00107759	963	927369	31.0322	.00103842	998	996004	31.5911	.00100200
929	863041	30.4795	.00107643	964	929296	31.0483	.00103734	999	998001	31.6070	.00100100
930	864900	30.4959	.00107527	965	931225	31.0644	.00103627	1000	1000000	31.6228	.00100000
931	866761	30.5123	.00107411	966	933156	31.0805	.00103520				
932	868624	30.5287	.00107296	967	935089	31.0966	.00103413				
933	870489	30.5450	.00107181	968	937024	31.1127	.00103306				
934	872356	30.5614	.00107066	969	938961	31.1288	.00103199				
935	874225	30.5778	.00106952	970	940900	31.1448	.00103093				

Answers

CHAPTER 1 EXERCISES

1. 1792   2. 19   3. -3 4. -256

5. 210   6. 0.64   7. -139 8. 69

9. 77   10. -16 11. 12   12. 56

13. 72   14. -30 15. -112 16. 3

17. -1 18. $4\,(1) + 5\,(6) + 2\,(7) + 3\,(8) = 72$ 19. $4\,(2) + 5\,(-3) + 3\,(4) = 5$

20. $a = 19 - b - c$ or $a = 19 - (b + c)$ 21. $y = 18 + n$ 22. $\Sigma X = N\bar{X}$

23. $N = \dfrac{\Sigma X}{X}$ 24. $n - 1 = \dfrac{\Sigma x^2}{s^2}$ 25. $\Sigma X = N\bar{X}$ 26. $N = \dfrac{\Sigma X}{\bar{X}}$

27. $\Sigma x^2 = (n - 1)\, s^2$ 28. $n - 1 = \dfrac{\Sigma x^2}{s^2}$ 29. $X^{a+b} = X^{10}$

30. $(6^5)(6^3) = 6^8 = 1{,}679{,}616$ 31. $(1/4)^2 \cdot (1/4)^4 = (1/4)^6 = 1/4096$

32. $\dfrac{X^y}{X^a} = X^{y-a}$ 33. $\dfrac{4^6}{4^3} = 4^3 = 64$

34. $\dfrac{2^2}{2^5} = 2^{-3} = (1/2)^3 = 1/8$ 35. $\dfrac{7^4}{7^7} = 7^{-3} = (1/7)^3 = 1/343$

36. $\dfrac{3^5}{3^5} = 3^0 = 1$ 37. $\sqrt{30.00} = 5.4772$

[Note: any value raised to the zero power is equal to 1. In this example,

$$\frac{3^5}{3^5} = 1.$$

Any number divided by itself is equal to 1. However:

$$\frac{3^5}{3^5} = 3^0,$$

therefore $3^0 = 1$.]

38. $\sqrt{300.00} = 17.3205$ 39. $\sqrt{3000.00} = 54.772$ 40. $\sqrt{30000.00} = 173.205$

41. $\sqrt{169} = 13.0000$ 42. $\sqrt{1600} = 40.00$ 43. 0.05

44. 78.4   45. 85.467   46. 0.06

47. 0.1   48. 649   49. 70.00

50. 70.8   51. 70.6   52. 70.650

CHAPTER 1 TEST

1. d   2. a   3. c   4. a   5. b

6. b   7. d   8. a   9. d   10. c

11. b	12. d	13. d	14. d	15. a
16. c	17. d	18. c	19. a	20. a
21. c	22. c	23. a	24. d	25. c
26. b	27. b	28. b	29. a	30. d

CHAPTER 2 EXERCISES

1. $\Sigma(X_5, X_7, X_{10})$ 2. $\Sigma(3, 7, 12, 15)$ 3. $\Sigma(A, C, Y)$ 4. $\sum_{i=5}^{N} X_i$ 5. $\sum_{i=1}^{6} X_i$

6. $\sum_{i=1}^{N} X_i$ 7. $\sum_{i=4}^{7} X_i$ 8. 36   9. 20   10. 32

11. 11   12. 22   13. 44   14. 76   15. 60

16. 20   17. -20 18. 4   19. Sex   20. Color

21. Zip code   22. Hair color   23. Religious preference   24. Occupation

25. Number of blonds in class

26. Number of Datsuns sold in Yasu City during 1976

27. The population of each of the original thirteen colonies at year's end in 1976

28. Rank in leadership qualities

29. Worse natural disaster in history

30. Post position in horse race   31. General > Private   32. Corporal < Sergeant

33. Leader > Follower   34. Third post position > eighth post position

35. Winner of beauty pageant > also ran

36. Player sent down to minor leagues < most valuable player on team

37. Cars sold in Yasu City by manufacturers

38. Number of horses during 1976 which finished first at each post position on an oval track

39. Number of different types of crimes during 1977   40. Height of tallest person in class

41. The weight of a Datsun B-210

42. The low temperature in Gunnison, Colorado on Jan. 5, 1977

43. The size of the family   44. D   45. C

46. D   47. C   48. Heights of various athletes

49. The number of children per family unit   50. Daily miles driven to work by commuters

51. Male = 0.77; female = 0.23

52. Guns = 0.66; cutting or stabbing = 0.17; strangulations or beatings = 0.08; blunt object = 0.05; arson = 0.01; other = 0.02

53. Suicides = 0.5445; homicides = 0.4555   54. Male = 77.40%; female = 22.60%

55. Guns = 65.70%; cutting or stabbing = 17.43%;

strangulations or beatings = 8.44%;

blunt object = 4.95%; arson = 1.01%; other = 2.47%

56. Suicides = 54%; homicides = 46%   57. 2.22:1   58. 2.08:1

59.

Year	Number of female victims of homicide	Time ratio (fixed-base 1969)
1968	1,700	94.39
1969	1,801	100.00
1970	1,938	107.61
1971	2,106	116.94
1972	2,156	119.71
1973	2,575	142.98

60.

Year	U.S. population (in thousands)	Time ratio (fixed-base 1969)
1968	199,312	99.01
1969	201,306	100.00
1970	203,806	101.24
1971	206,212	102.44
1972	208,230	103.44
1973	209,844	104.24

61.

Year	Number of female victims of homicide	Moving-base time ratio (preceding year)
1968	1,700	———
1969	1,801	105.94
1970	1,938	107.61
1971	2,106	108.67
1972	2,156	102.37
1973	2,575	119.43

62.

Year	U.S. population (in thousands)	Moving-base time ratio (preceding year)
1968	199,312	———
1969	201,306	101.00
1970	203,806	101.24
1971	206,212	101.18
1972	208,230	100.98
1973	209,844	100.78

CHAPTER 2 TEST

1. c	2. c	3. d	4. a	5. d
6. a	7. d	8. b	9. d	10. a
11. b	12. b	13. a	14. c	15. c
16. d	17. d	18. b	19. d	20. c
21. b	22. c	23. a	24. d	25. a
26. c	27. b	28. b	29. a	30. d

31.

	Fixed-base time ratios (base year 1969)	
Year	White	Nonwhite
1968	97.54	88.47
1969	100.00	100.00
1970	104.74	106.90
1971	106.15	108.96
1972	110.68	133.06
1973	113.01	132.34

32.

	Moving-base time ratios (base: preceding year)	
Year	White	Nonwhite
1968	———	———
1969	102.52	113.04
1970	104.74	106.90
1971	101.35	101.93
1972	104.27	122.12
1973	102.11	99.46

33.

Year	White male
1968	15.43
1969	15.82
1970	16.57
1971	16.79
1972	17.51
1973	17.88

34. Year	Nonwhite male
1968	.1321
1969	.1493
1970	.1596
1971	.1627
1972	.1987
1973	.1976

35. Year	Interclass ratio
1968	16.90:1
1969	15.33:1
1970	15.02:1
1971	14.94:1
1972	12.75:1
1973	13.09:1

CHAPTER 3 EXERCISES

1.

MPG	MPG
28	18
27	17
26	16
25	15
24	14
23	13
22	12
21	11
20	10
19	

2.

MPG	f	MPG	f
28	1	18	
27		17	11
26		16	111
25		15	111
24	1	14	
23		13	111
22		12	11111
21	11	11	11111
20		10	11
19	111		

3.

Class interval	f
28–29	1
26–27	0
24–25	1
22–23	0
20–21	2
18–19	3
16–17	5
14–15	3
12–13	8
10–11	7

4.

Class interval	f
28–30	1
25–27	0
22–24	1
19–21	5
16–18	5
13–15	6
10–12	12

5.

Class interval	f
25–29	1
20–24	3
15–19	11
10–14	15

6. a) 15.5–16.6
   b) 11.5–12.5
   c) −0.5–0.5
   d) 99.5–100.5

7. a) 155–165
   b) 185–195
   c) 190–200
   d) 865–875

8. a) 65.45–65.55
   b) 192.95–193.05
   c) 100.05–100.15
   d) 99.85–99.95

9. 9.5

10. 18.5

11. 25.5

12. 4.5

13. −0.5

14. 15.5

15. 99.5

16. 59.5

17. 12.5

18. 21.5

19. 28.5

20. 9.5

21. 4.5

22. 23.5

23. 110.5

24. 69.5

25. Class interval

Apparent limits	Real limits	f	Cum f
28–29	27.5–29.5	1	30
26–27	25.5–27.5	0	29
24–25	23.5–25.5	1	29
22–23	21.5–23.5	0	28
20–21	19.5–21.5	2	28
18–19	17.5–19.5	3	26
16–17	15.5–17.5	5	23
14–15	13.5–15.5	3	18
12–13	11.5–13.5	8	15
10–11	9.5–11.5	7	7

26. Class interval

Apparent limits	Real limits	f	Cum f
28–30	27.5–30.5	1	30
25–27	24.5–27.5	0	29
22–24	21.5–24.5	1	29
19–21	18.5–21.5	5	28
16–18	15.5–18.5	5	23
13–15	12.5–15.5	6	18
10–12	9.5–12.5	12	12

27.

Class interval			
Apparent limits	Real limits	f	Cum f
25–29	24.5–29.5	1	30
20–24	19.5–24.5	3	29
15–19	14.5–19.5	11	26
10–14	9.5–14.5	15	15

28. 25.5   29. 15.5   30. 13.5

31. 21.5   32. 17.5

33.

Class interval				
Apparent limits	Real limits	f	Cum f	Cum %
28–29	27.5–29.5	1	30	100
26–27	25.5–27.5	0	29	97
24–25	23.5–25.5	1	29	97
22–23	21.5–23.5	0	28	93
20–21	19.5–21.5	2	28	93
18–19	17.5–19.5	3	26	87
16–17	15.5–17.5	5	23	77
14–15	13.5–15.5	3	18	60
12–13	11.5–13.5	8	15	50
10–11	9.5–11.5	7	7	23
		$N = 30$		

34.

Class interval				
Apparent limits	Real limits	f	Cum f	Cum %
28–30	27.5–30.5	1	30	100
25–27	24.5–27.5	0	29	97
22–24	21.5–24.5	1	29	97
19–21	18.5–21.5	5	28	93
16–18	15.5–18.5	5	23	77
13–15	12.5–15.5	6	18	60
10–12	9.5–12.5	12	12	40

35.

Class interval				
Apparent limits	Real limits	f	Cum f	Cum %
25–29	24.5–29.5	1	30	100
20–24	19.5–24.5	3	29	97
15–19	14.5–19.4	11	26	87
10–14	9.5–14.5	5	15	50

36. 88% above; 12% below

37. 97% above; 3% below

38. 40% above; 60% below

39. 23% above; 77% below

40. 0% above; 100% below

41. 67% above; 33% below

CHAPTER 3 TEST

1. b   2. a   3. c   4. c   5. d

6. d   7. b   8. b   9. b   10. d

11. a   12. c   13. c   14. a   15. d

16. b   17. c   18. c   19. a   20. c

21. b   22. d   23. a   24. b   25. c

26. b

27.

X	f	X	f	X	f	X	f	X	f
4.6	1	3.9	1	3.2	2	2.5	3	1.8	2
4.5	1	3.8	0	3.1	1	2.4	0	1.7	1
4.4	1	3.7	2	3.0	3	2.3	2	1.6	1
4.3	1	3.6	1	2.9	4	2.2	1	1.5	1
4.2	1	3.5	2	2.8	6	2.1	1	1.4	1
4.1	0	3.4	3	2.7	3	2.0	2	1.3	0
4.0	1	3.3	1	2.6	3	1.9	0	1.2	1

28.

Class interval	f
4.5–4.7	3
4.2–4.4	2
3.9–4.1	2
3.6–3.8	3
3.3–3.5	6
3.0–3.2	6
2.7–2.9	13
2.4–2.6	6
2.1–2.3	4
1.8–2.0	4
1.5–1.7	3
1.2–1.4	2

29.

Apparent limits	Real limits	f	Cum f	Cum %
4.5–4.7	4.45–4.75	3	54	100.00
4.2–4.4	4.15–4.45	2	51	94.44
3.9–4.1	3.85–4.15	2	49	90.74
3.6–3.8	3.55–3.85	3	47	87.04
3.3–3.5	3.25–3.55	6	44	81.48
3.0–3.2	2.95–3.25	6	38	70.37
2.7–2.9	2.65–2.95	13	32	59.26
2.4–2.6	2.35–2.65	6	19	35.19
2.1–2.3	2.05–2.35	4	13	24.07
1.8–2.0	1.75–2.05	4	9	16.67
1.5–1.7	1.45–1.75	3	5	9.26
1.2–1.4	1.15–1.45	2	2	3.70
		$N = 54$		

30. a) 2.95   b) 3.85   c) 1.75   d) 2.05

31. a) 59.26   b) 3.70   c) 70.37   d) 90.74

CHAPTER 4 EXERCISES

1.

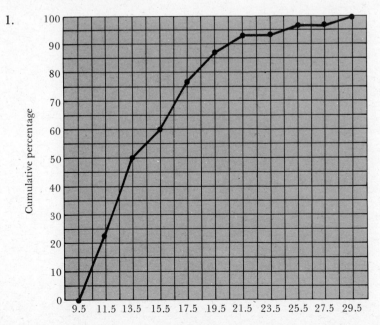

2.

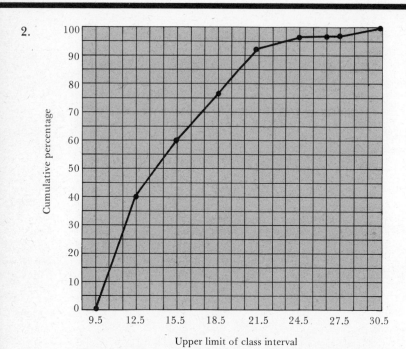

3.

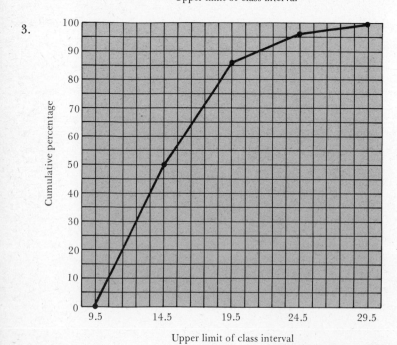

Answers to Exercises 4 through 17 are only approximate, based on reading a graph. If your answer is within two percentage points of those shown below, consider your answer correct.

4. 18	5. 42	6. 48	7. 63	8. 70
9. 74	10. 80	11. 90	12. 92	13. 94
14. 96	15. 97	16. 98	17. 100	

Answers to Exercises 18 through 30 are only approximate, based on reading a graph. If your answer is within one score unit of those shown below, consider your answer correct.

18. 10	19. 10.5	20. 11	21. 12	22. 13
23. 14	24. 15	25. 16	26. 17	27. 18
28. 20	29. 22	30. 25	31. 89.58	32. 54.17
33. 94.17	34. 74.17	35. 17.50	36. 79.58	37. 10.36
38. 19.27	39. 11.70	40. 11.04	41. 20.76	42. 10.87

CHAPTER 4 TEST

1. b	2. a	3. b	4. d	5. a
6. b	7. c	8. b	9. c	10. d
11. c	12. a	13. b	14. d	15. c
16. a	17. d	18. c	19. b	20. d

21.

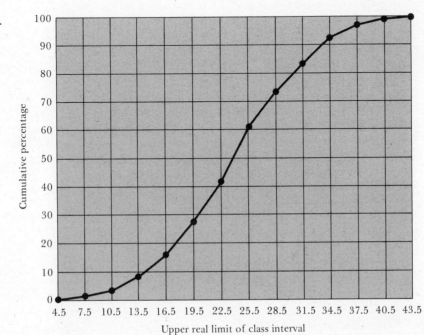

22. Any answer included in the following range is acceptable.

    a) 5–7   b) 50–52   c) 84–86   d) 92–94   e) 98–100

23. Any answer included in the following range is acceptable.

    a) 17–29   b) 20–22   c) 24–26   d) 29–31   e) 34–36

24. a) 5.61   b) 51.02   c) 85.20   d) 93.19   e) 99.15

25. a) 17.48   b) 21.27   c) 24.49   d) 29.83   e) 35.34

CHAPTER 5 EXERCISES

1. $\bar{X} = 14.07$ 2. Median = 14   3. Mode = 11

4.

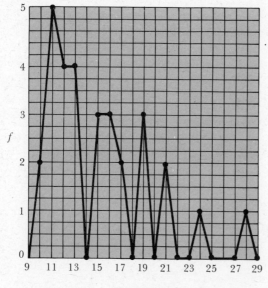

Number of cars obtaining f indicated mpg

Miles per gallon

5. $\bar{X} = 14.97$ 6. $\bar{X} = 15.95$ 7. $\bar{X} = 13.00$ 8. $\bar{X}_T = 13.67$ 9. $\bar{X}_T = 79.55$

10. $\Sigma X = 45$ 11. $\Sigma X = 343$ 12. $\Sigma X = 1284$ 13. $\Sigma X = 2232$ 14. $\bar{X}_T = 47.66$

15. $\bar{X}_T = 9.74$ 16. $\bar{X}_T = 8.88$ 17. $\bar{X}_T = 16.49$

18.

Real limits of interval	f	Cum f
27.5–30.5	1	60
24.5–27.5	1	59
21.5–24.5	5	58
18.5–21.5	11	53
15.5–18.5	12	42
12.5–15.5	11	30
9.5–12.5	19	19

Median = 15.5

19.

Real limits of interval	f	Cum f
27.5–29.5	1	30
25.5–27.5	0	29
23.5–25.5	1	29
21.5–23.5	0	28
19.5–21.5	2	28
17.5–19.5	3	26
15.5–17.5	5	23
13.5–15.5	3	18
11.5–13.5	8	15
9.5–11.5	7	7

Median = 13.5

20.

Midpoint of interval	f	Cum f
23.5–26.5	1	46
20.5–23.5	2	45
17.5–20.5	4	43
14.5–17.5	8	39
11.5–14.5	16	31
8.5–11.5	8	15
5.5– 8.5	4	7
2.5– 5.5	2	3
−0.5– 2.5	1	1

Median = 13.00

21. Mode = 12.5   22. Mode = 11   23. Mode = 13   24. Identical

CHAPTER 5 TEST

1. c   2. a   3. d   4. b   5. b
6. c   7. a   8. d   9. d   10. b
11. d   12. d   13. c   14. a   15. b

16.

Class interval	f
42–45	1
38–41	5
34–37	2
30–33	3
26–29	5
22–25	15
18–21	5
14–17	4
10–13	2
6– 9	3
2– 5	3
	$N = 48$

17. $\bar{X} = 23.17$; median = 23.4; mode = 23.5

18. a) $\Sigma X_T = 465.9$, $N_T = 46$, $\bar{X}_T = 10.13$

  b) $\Sigma X_T = 13564$, $N_T = 179$, $\bar{X}_T = 75.78$

  c) $\Sigma X_T = 724$, $N_T = 60$, $\bar{X}_T = 12.07$

CHAPTER 6 EXERCISES

1. b   2. a   3. a   4. a   5. b
6. 24   7. 78   8. 17   9. 1364   10. 104
11. 41   12. 321

13.

Real limits of interval	f	Cum f
27.5–30.5	1	60
24.5–27.5	1	59
21.5–24.5	5	58
18.5–21.5	11	53
15.5–18.5	12	42
12.5–15.5	11	30
9.5–12.5	19	19

Interquartile range = 11.87–19.32

14.

Real limits of interval	f	Cum f
27.5–29.5	1	30
25.5–27.5	0	29
23.5–25.5	1	29
21.5–23.5	0	28
19.5–21.5	2	28
17.5–19.5	3	26
15.5–17.5	5	23
13.5–15.5	3	18
11.5–13.5	8	15
9.5–11.5	7	7

Interquartile range: 11.62–17.30

15. $(105 - 110) + (110 - 110) + (115 - 110) = -5 + 0 + 5 = 0$

16.

X	$X - \bar{X}$
20	−10
25	−5
25	−5
30	0
30	0
30	0
35	+5
35	+5
40	+10
$\bar{X} = 30$	$\Sigma(X - \bar{X}) = 0$

17. $|-24| = 24$

18. $|15| = 15$

19. $|-1| = 1$

20. $|0| = 0$

21. $|-453| = 453$

22. $|1| = 1$

23. $|-0.06| = 0.06$

24.

X	$|X - \bar{X}|$
115	$|5|$
110	$|0|$
105	$|-5|$
	$\Sigma(|X - \bar{X}|) = 10$

M.D. $= \dfrac{10}{3} = 3.33$

25.

X	$|X - \bar{X}|$
40	$|10|$
35	$|5|$
35	$|5|$
30	$|0|$
30	$|0|$
30	$|0|$
25	$|-5|$
25	$|-5|$
20	$|-10|$
	$\Sigma(|X - \bar{X}|) = 40$

M.D. $= \dfrac{40}{9} = 4.44$

26.

X	$|X - \bar{X}|$
15	$|5|$
13	$|3|$
12	$|2|$
10	$|0|$
8	$|-2|$
7	$|-3|$
5	$|-5|$
$\bar{X} = 10$	$\Sigma(|X - \bar{X}|) = 20$

M.D. $= \dfrac{20}{7} = 2.86$

27.

X	f	$|X - \bar{X}|$	$f(|X - \bar{X}|)$
40	1	$|10|$	10
35	2	$|5|$	10
30	3	$|0|$	0
25	2	$|-5|$	10
20	1	$|-10|$	10
$\bar{X} = 30$	$N = 9$		$\Sigma f(|X - \bar{X}|) = 40$

M.D. $= \dfrac{40}{9} = 4.44$

28.

Class interval	Midpoint of interval	f	$\lvert X - \bar{X}\rvert$	$f(\lvert X - \bar{X}\rvert)$
16–18	17	2	$\lvert 7.5\rvert$	15.0
13–15	14	5	$\lvert 4.5\rvert$	22.5
10–12	11	9	$\lvert 1.5\rvert$	13.5
7– 9	8	9	$\lvert -1.5\rvert$	13.5
4– 6	5	5	$\lvert -4.5\rvert$	22.5
1– 3	2	2	$\lvert -7.5\rvert$	15.0
$\bar{X} = 9.5$		$N = 32$		$\Sigma f(\lvert X - \bar{X}\rvert) = 102$

$$\text{M.D.} = \frac{102}{32} = 3.19$$

These answers to Exercises 29 through 38 correspond to the section in which s^2 and s are defined by $N - 1$ in the denominator. If you used N in the denominator, go forward to answers 29A through 38A.

29.

X	$X - \bar{X}$	$(X - \bar{X})^2$
105	−7.5	56.25
110	−2.5	6.25
115	2.5	6.25
120	7.5	56.25
$\Sigma X = 450$		$\Sigma(X - \bar{X})^2 = 125$

$\bar{X} = 112.5$

$$s^2 = \frac{125}{3} = 41.67$$

$s = 6.46$

30.

X	$X - \bar{X}$	$(X - \bar{X})^2$
5	−7.5	56.25
10	−2.5	6.25
15	2.5	6.25
20	7.5	56.25
$\Sigma X = 50$		$\Sigma(X - \bar{X})^2 = 125$

$\bar{X} = 12.5$

$$s^2 = \frac{125}{3} = 41.67$$

$s = 6.46$

31.

X	$X - \bar{X}$	$(X - \bar{X})^2$
27	−7.5	56.25
32	−2.5	6.25
37	2.5	6.25
42	7.5	56.25
$\Sigma X = 138$		$\Sigma(X - \bar{X})^2 = 125.00$

$\bar{X} = 34.5$

$s^2 = 41.67$

$s = 6.46$

32.

X	$X - \bar{X}$	$(X - \bar{X})^2$
1	−9	81
8	−2	4
15	5	25
16	6	36
$\Sigma X = 40$		$\Sigma(X - \bar{X})^2 = 146$

$\bar{X} = 10$

$s^2 = 48.67$

$s = 6.98$

33.

X	X^2
20	400
15	225
10	100
5	25
$\Sigma X = 50$	$\Sigma X^2 = 750$

$$\Sigma(X - \bar{X})^2 = 750 - \frac{(50)^2}{4}$$

$$= 750 - 625$$

$$= 125$$

$$s^2 = \frac{125}{3} = 41.67$$

$$s = \sqrt{41.67} = 6.46$$

34.

X	X^2
23	529
17	289
15	225
14	196
11	121
9	81
5	25
$\Sigma X = 94$	$\Sigma X^2 = 1466$

$$\Sigma(X - \bar{X})^2 = 1466 - \frac{(94)^2}{7}$$

$$= 1466 - 1262.29$$

$$= 203.71$$

$$s^2 = 33.95$$

$$s = 5.83$$

35.

X	X^2
16.5	272.25
13.2	174.24
11.6	134.56
10.4	108.16
9.3	86.49
7.1	50.41
4.5	20.25
$\Sigma X = 72.6$	$\Sigma X^2 = 846.36$

$$\Sigma(X - \bar{X})^2 = 846.36 - \frac{(72.6)^2}{7}$$

$$= 93.39$$

$$s^2 = 15.57$$

$$s = 3.95$$

36.

X	f	fX	X^2	fX^2
15	1	15	225	225
14	4	56	196	784
13	8	104	169	1352
12	16	192	144	2304
11	2	22	121	242
10	12	120	100	1200
9	1	9	81	81
$N = 44$		$\Sigma fX = 518$		$\Sigma fX^2 = 6188$

$$\Sigma f(X - \bar{X})^2 = 6188 - \frac{(518)^2}{44}$$

$$= 6188 - 6098.27$$

$$= 89.73$$

$$s^2 = 2.09$$

$$s = 1.45$$

37.

Class interval	Midpoint of interval	f	fX	X^2	fX^2
28–30	29	1	29	841	841
25–27	26	1	26	676	676
22–24	23	5	115	529	2645
19–21	20	11	220	400	4400
16–18	17	12	204	289	3468
13–15	14	11	154	196	2156
10–12	11	19	209	121	2299
		$N = 60$	$\Sigma fX = 957$		$\Sigma fX^2 = 16485$

$$\Sigma f(X - \bar{X})^2 = 16485 - \frac{(957)^2}{60}$$

$$= 16485 - 15264.15$$

$$= 1220.85$$

$$s^2 = 20.69$$

$$s = 4.55$$

38.

Class interval	Midpoint of interval	f	fX	X^2	fX^2
28–29	28.5	1	28.5	812.25	812.25
26–27	26.5	0	00.0	702.25	0.00
24–25	24.5	1	24.5	600.25	600.25
22–23	22.5	0	00.0	506.25	0.00
20–21	20.5	2	41.0	420.25	840.50
18–19	18.5	3	55.5	342.25	1026.75
16–17	16.5	5	82.5	272.25	1361.25
14–15	14.5	3	43.5	210.25	630.75
12–13	12.5	8	100.0	156.25	1250.00
10–11	10.5	7	73.5	110.25	771.75
		$N = 30$	$\Sigma fX = 449$		$\Sigma fX^2 = 7293.50$

$$\Sigma f(X - \bar{X})^2 = 7293.50 - \frac{(449)^2}{30}$$

$$= 7293.50 - 6720.03$$

$$= 573.47$$

$$s^2 = 19.77$$

$$s = 4.45$$

These answers to Exercises 29A through 38A correspond to the section in which s^2 and s are defined by N in the denominator. If you used $N - 1$ in the denominator, go back to answers 29 through 38.

29A.

X	$X - \bar{X}$	$(X - \bar{X})^2$
105	−7.5	56.25
110	−2.5	6.25
115	2.5	6.25
120	7.5	56.25
$\Sigma X = 450$		$\Sigma(X - \bar{X})^2 = 125$

$\bar{X} = 112.5$

$$s^2 = \frac{125}{4} = 31.25$$

$$s = 5.59$$

30A.

X	$X - \bar{X}$	$(X - \bar{X})^2$
5	−7.5	56.25
10	−2.5	6.25
15	2.5	6.25
20	7.5	56.25
$\Sigma X = 50$		$\Sigma(X - \bar{X})^2 = 125$

$\bar{X} = 12.5$

$$s^2 = \frac{125}{4} = 31.25$$

$$s = 5.59$$

31A.

X	$X - \bar{X}$	$(X - \bar{X})^2$
27	−7.5	56.25
32	−2.5	6.25
37	2.5	6.25
42	7.5	56.25
$\Sigma X = 138$		$\Sigma(X - \bar{X})^2 = 125.00$

$\bar{X} = 34.5$

$$s^2 = \frac{125}{4} = 31.25$$

$$s = 5.59$$

32A.

X	$X - \bar{X}$	$(X - \bar{X})^2$
1	−9	81
8	−2	4
15	5	25
16	6	36
$\Sigma X = 40$		$\Sigma(X - \bar{X})^2 = 146$

$\bar{X} = 10$

$$s^2 = \frac{146}{4} = 36.50$$

$$s = 6.04$$

33A.

X	X^2
20	400
15	225
10	100
5	25
$\Sigma X = 50$	$\Sigma X^2 = 750$

$$\Sigma(X - \bar{X})^2 = 750 - \frac{(50)^2}{4}$$

$$= 750 - 625$$

$$= 125$$

$$s^2 = \frac{125}{4} = 36.50$$

$$s = \sqrt{36.50} = 6.04$$

34A.

X	X^2
23	529
17	289
15	225
14	196
11	121
9	81
5	25
$\Sigma X = 94$	$\Sigma X^2 = 1466$

$$\Sigma(X - \bar{X})^2 = 1466 - \frac{(94)^2}{7}$$

$$= 1466 - 1262.29$$

$$= 203.71$$

$$s^2 = 29.10$$

$$s = 5.39$$

35A.

X	X^2
16.5	272.25
13.2	174.24
11.6	134.56
10.4	108.16
9.3	86.49
7.1	50.41
4.5	20.25
$\Sigma X = 72.6$	$\Sigma X^2 = 846.36$

$$\Sigma(X - \bar{X})^2 = 846.36 - \frac{(72.6)^2}{7}$$

$$= 93.39$$

$$s^2 = 13.34$$

$$s = 3.65$$

36A.

X	f	fX	X^2	fX^2
15	1	15	225	225
14	4	56	196	784
13	8	104	169	1352
12	16	192	144	2304
11	2	22	121	242
10	12	120	100	1200
9	1	9	81	81
	$N = 44$	$\Sigma fX = 518$		$\Sigma fX^2 = 6188$

$$\Sigma f(X - \bar{X})^2 = 6188 - \frac{(518)^2}{44}$$

$$= 6188 - 6098.27$$

$$= 89.73$$

$$s^2 = 2.04$$

$$s = 1.43$$

37A.

Class interval	Midpoint of interval	f	fX	X^2	fX^2
28–30	29	1	29	841	841
25–27	26	1	26	676	676
22–24	23	5	115	529	2645
19–21	20	11	220	400	4400
16–18	17	12	204	289	3468
13–15	14	11	154	196	2156
10–12	11	19	209	121	2299
		$N = 60$	$\Sigma fX = 957$		$\Sigma fX^2 = 16485$

$$\Sigma f(X - \bar{X})^2 = 16485 - \frac{(957)^2}{60}$$

$$= 16485 - 15264.15$$

$$= 1220.85$$

$$s^2 = 20.35$$

$$s = 4.51$$

38A.

Class interval	Midpoint of interval	f	fX	X^2	fX^2
28–29	28.5	1	28.5	812.25	812.25
26–27	26.5	0	00.0	702.25	0.00
24–25	24.5	1	24.5	600.25	600.25
22–23	22.5	0	00.0	506.25	0.00
20–21	20.5	2	41.0	420.25	840.50
18–19	18.5	3	55.5	342.25	1026.75
16–17	16.5	5	82.5	272.25	1361.25
14–15	14.5	3	43.5	210.25	630.75
12–13	12.5	8	100.0	156.25	1250.00
10–11	10.5	7	73.5	110.25	771.75
		$N = 30$	$\Sigma fX = 449$		$\Sigma fX^2 = 7293.50$

$$\Sigma f(X - \bar{X})^2 = 7293.50 - \frac{(449)^2}{30}$$

$$= 7293.50 - 6720.03$$

$$= 573.47$$

$$s^2 = 19.12$$

$$s = 4.37$$

CHAPTER 6 TEST

The following answers are correct if you defined s and s^2 by the use of $N - 1$ in the denominator.

1. d	2. c	3. b	4. d	5. a
6. a	7. c	8. b	9. a	10. d
11. c	12. b	13. a	14. b	15. d

16. $\Sigma X = 896$

    $N = 49$

    $\Sigma X^2 = 16478$

    $\Sigma(X - \bar{X})^2 = 94$

    $\bar{X} = 18.29$

    $s^2 = 1.96$

    $s = 1.40$

17. $\Sigma X = 948$

    $N = 40$

    $\Sigma X^2 = 26050$

    $\Sigma(X - \bar{X})^2 = 3582.4$

    $\bar{X} = 23.70$

    $s^2 = 91.86$

    $s = 9.58$

The following answers are correct if you defined s and s^2 by the use of N in the denominator.

1. d	2. c	3. b	4. d	5. a
6. c	7. b	8. b	9. a	10. d
11. a	12. d	13. d	14. a	15. b

16. $\Sigma X = 896$

$N = 49$

$\Sigma X^2 = 16478$

$\Sigma(X - \bar{X})^2 = 94$

$\bar{X} = 18.29$

$s^2 = 1.92$

$s = 1.39$

17. $\Sigma X = 948$

$N = 40$

$\Sigma X^2 = 26050$

$\Sigma(X - \bar{X})^2 = 3582.4$

$\bar{X} = 23.70$

$s^2 = 89.56$

$s = 9.46$

CHAPTER 7 EXERCISES

1. Normal distribution

2. Nonnormal distribution, positively skewed

3. Multimodal distribution

4. Normal distribution

5. Nonnormal distribution, negatively skewed

6. 34.13%

7. 13.59%

8. 50%

9. 84%

10. 15.74%

11. 15.74%

12. 47.72%

13. $z = 2.00$

14. $z = -1.00$

15. $z = 0.00$

16. $z = 3.00$

17. $z = -3.00$

18. 1

19. 7

20. 8

21. 21

22. 42

23. 54

24. 79

25. 96

26. 97

27. 99

28. $z = -0.9$; percentile rank = 18

29. $z = 1.6$; percentile rank = 95

30. $z = -1.6$; percentile rank = 5

31. $z = 0.5$; percentile rank = 69

32. $z = -0.4$; percentile rank = 34

33. $z = 1.1$; percentile rank = 86

34. $z = -0.7$; percentile rank = 24

35. $z = 2.2$; percentile rank = 99

36. $z = 0.17$; percentile rank = 56.75

37. $z = 1.33$; percentile rank = 90.82

38. $z = -1.83$; percentile rank = 3.36

39. $z = -2.67$; percentile rank = 00.38

40. $z_{x_2} = -0.38$; percentile rank = 35.20
$z_{x_1} = -0.69$; percentile rank = $\underline{24.51}$
$10.69\,\%$

41. $z_{x_2} = -1.31$; percentile rank = 9.51
$z_{x_1} = -1.44$; percentile rank = $\underline{7.49}$
$2.02\,\%$

42. $z_{x_2} = -0.12$; percentile rank = 45.22
$z_{x_1} = -0.25$; percentile rank = $\underline{40.13}$
$5.09\,\%$

43. $z_{x_1} = -0.06$; percentile rank = 47.61
$z_{x_2} = -2.19$; percentile rank = $\underline{01.43}$
$46.18\,\%$

44. $z_{x_1} = 2.25$; percentile rank = 98.78
$z_{x_2} = 1.00$; percentile rank = $\underline{84.13}$
$14.65\,\%$

45. $z_{x_1} = 0.31$; percentile rank = 62.17
$z_{x_2} = 0.06$; percentile rank = $\underline{52.39}$
$9.78\,\%$

46. $z_{x_1} = 2.31$; percentile rank = 98.96
$z_{x_2} = 0.25$; percentile rank = $\underline{59.87}$
$39.09\,\%$

47. $z_{x_1} = 1.06$; percentile rank = 85.54
$z_{x_2} = 0.62$; percentile rank = $\underline{73.24}$
$12.30\,\%$

48. $z_{x_1} = 1.50$; percentile rank = 93.32
$z_{x_2} = -0.44$; percentile rank = $\underline{33.00}$
$60.32\,\%$

49. $z_{x_1} = 0.19$; percentile rank = 57.53
$z_{x_2} = -0.19$; percentile rank = $\underline{42.47}$
$15.06\,\%$

50. $z_{x_1} = 1.06$; percentile rank = 85.54
    $z_{x_2} = -1.00$; percentile rank = 15.87
    $\underline{}$
     69.67 %

51. $z_x = -0.50$; percentile rank = 30.85
    $z_y = -1.00$; percentile rank = 15.87

    Variable X is higher.

52. $z_x = 1.30$; percentile rank = 90.32
    $z_y = 1.67$; percentile rank = 95.25

    Variable Y is higher.

53. $z_x = -2.00$; percentile rank = 2.28
    $z_y = -2.00$; percentile rank = 2.28

    The percentile rank of each score is identical.

54. $z_x = 0.20$; percentile rank = 57.93
    $z_y = 0.67$; percentile rank = 74.86

    Variable Y is higher.

55. $z_x = -0.20$; percentile rank = 42.07
    $z_y = -0.67$; percentile rank = 25.14

    Variable X is higher.

56. $V = 9.23$ 57. $V = 2.73$ 58. $V = 1.12$ 59. $V = 1.33$ 60. $V = 2.50$

CHAPTER 7 TEST

1. b   2. a   3. d   4. b   5. d
6. a   7. d   8. b   9. d   10. c
11. a   12. b   13. d   14. b   15. a
16. c   17. c   18. c   19. c   20. c

21.

	z	Percentile rank
a)	-0.67	25.14
b)	0.56	71.23
c)	1.33	90.82
d)	2.50	99.38
e)	-1.33	09.18
f)	2.17	98.50
g)	0.13	56.17
h)	0.20	57.93
i)	-1.47	07.08

22. a) F ($z = -2.00$) vs. E ($z = -2.50$)
    b) E ($z = 3.12$) vs. F ($z = 2.20$)
    c) F ($z = 2.50$) vs. E ($z = 1.67$)
    d) F ($z = -1.20$) vs. E ($z = -1.25$)

23. a) $V = 13.28$
    b) $V = 17.02$
    c) $V = 19.73$
    d) $V = 14.03$

CHAPTER 8 EXERCISES

	(A) Score	(B) Percentile Rank
1.	27.4	
	27	
	26	
	25	
	24	
	23	
	22	
	21	
	20	
	19	
	18	
	17	
	16	
	15	
	14	
Mean	13.81	50
	13	
	12	
	11	
	10	
	9	
	8	
	7	
	6	
	5	
	4	
	3	
	2	
	1	
	0.6	

	Percentile Rank
2.	99.95
3.	99.70
4.	99.20
5.	98.68
6.	97.62
7.	96.03
8.	94.44
9.	91.54
10.	87.30
11.	83.06
12.	78.05
13.	72.22
14.	66.40
15.	59.25
16.	50.79
17.	50.00
18.	42.33
19.	35.44
20.	30.16
21.	24.87
22.	19.84
23.	15.08
24.	10.32
25.	06.87
26.	04.76
27.	02.65
28.	01.32
29.	00.79
30.	00.27
31.	00.05

32. $z = 3.30$
33. $z = 2.75$
34. $z = 2.41$
35. $z = 2.22$
36. $z = 1.98$
37. $z = 1.75$
38. $z = 1.59$
39. $z = 1.37$
40. $z = 1.14$
41. $z = 0.96$
42. $z = 0.77$
43. $z = 0.59$
44. $z = 0.42$
45. $z = 0.23$
46. $z = 0.02$
47. $z = 0.00$
48. $z = -0.19$
49. $z = -0.37$
50. $z = -0.52$
51. $z = -0.68$
52. $z = -0.85$
53. $z = -1.03$
54. $z = -1.26$
55. $z = -1.49$
56. $z = -1.67$
57. $z = -1.94$
58. $z = -2.22$
59. $z = -2.42$
60. $z = -2.78$
61. $z = -3.30$

62. $T = 70 + 15z$
63. $T = 50 + 5z$
64. $T = 100 + 16z$
65. $T = 500 + 50z$
66. $T = 10 + 2z$
67. $T = 830$
68. $T = 775$
69. $T = 741$
70. $T = 722$
71. $T = 698$
72. $T = 675$
73. $T = 659$
74. $T = 637$
75. $T = 614$
76. $T = 596$
77. $T = 577$

78. $T = 559$
79. $T = 542$
80. $T = 523$
81. $T = 502$
82. $T = 500$
83. $T = 481$
84. $T = 463$
85. $T = 448$
86. $T = 432$
87. $T = 415$
88. $T = 397$
89. $T = 374$
90. $T = 351$
91. $T = 333$
92. $T = 306$
93. $T = 278$

94. $T = 258$
95. $T = 222$
96. $T = 170$

97. $z = \dfrac{300 - 500}{100} = -2.00$

Percentile rank = 2.28

98. $z = \dfrac{658 - 500}{100} = \dfrac{158}{100} = 1.58$

Percentile rank = 94.29

99. $z = \dfrac{260 - 500}{100} = -2.40$

Percentile rank = 0.82

100. $z = \dfrac{790 - 500}{100} = 2.90$

Percentile rank = 99.81

101. $z = \dfrac{480 - 500}{100} = 0.20$

Percentile rank = 57.92

102. 0.0017
103. 0.0091
104. 0.0219
105. 0.0339
106. 0.0562

107. 0.0863
108. 0.1127
109. 0.1561
110. 0.2083
111. 0.2516
112. 0.2966
113. 0.3352
114. 0.3653
115. 0.3885
116. 0.3989
117. 0.3989
118. 0.3918
119. 0.3725
120. 0.3485

121. 0.3166
122. 0.2780
123. 0.2347
124. 0.1804
125. 0.1315
126. 0.0989
127. 0.0608
128. 0.0339
129. 0.0213
130. 0.0084
131. 0.0017

132.

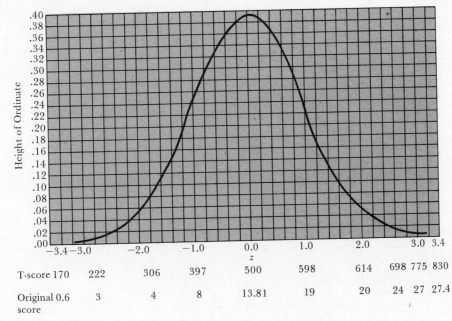

	z = −3.4	−3.0	−2.0	−1.0	0.0	1.0	2.0	3.0	3.4
T-score	170	222	306	397	500	598	614	698	775 830
Original score	0.6	3	4	8	13.81	19	20	24	27 27.4

CHAPTER 8 TEST

1. c
2. a
3. d
4. c
5. b
6. a
7. b
8. a
9. d
10. b
11. d
12. c
13. a
14. c
15. b

16.–17.

X	Percentile rank	Corresponding z	T	Height of ordinate
47.4	99.94	3.24	624	.0021
47.0	99.69	2.74	574	.0093
45.0	98.44	2.16	516	.0387
43.0	96.88	1.86	486	.0707
41.0	94.38	1.59	459	.1127
39.0	91.88	1.40	440	.1497
37.0	89.38	1.25	425	.1826
35.0	86.25	1.09	409	.2203
33.0	81.25	0.89	389	.2685
31.0	76.25	0.71	371	.3101
29.0	71.25	0.56	356	.3410
27.0	65.94	0.41	341	.3668
25.0	59.69	0.25	325	.3867
23.0	52.50	0.06	306	.3982
22.5	50	0.00	300	.3989
21.0	42.5	−0.19	281	.3918
19.0	33.44	−0.43	257	.3637
17.0	27.19	−0.61	239	.3312
15.0	20.94	−0.81	219	.2874
13.0	14.69	−1.05	195	.2299
11.0	8.75	−1.36	164	.1582
9.0	3.75	−1.78	122	.0818
7.6	0.25	−2.81	19	.0077

18.

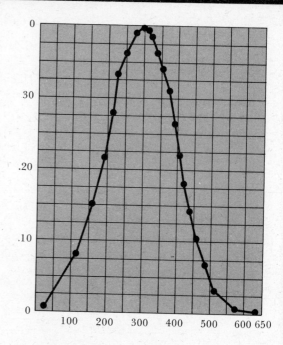

CHAPTER 9 EXERCISES

1.

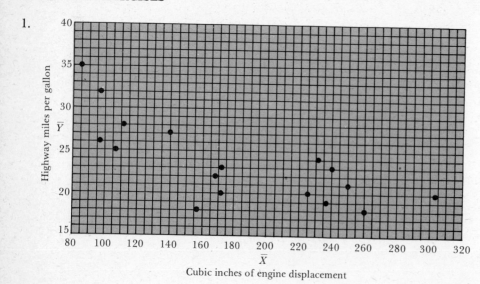

2. $\Sigma X = 54 \qquad \Sigma X^2 = 336 \qquad N = 9$

$$\frac{(\Sigma X)^2}{N} = 324$$

$$\Sigma x^2 = 336 - 324 = 12$$

3. $\Sigma Y = 189 \qquad \Sigma Y^2 = 4017 \qquad N = 9$

$$\frac{(\Sigma Y)^2}{N} = 3969$$

$$\Sigma y^2 = 4017 - 3969 = 88$$

4.

XY
100
115
115
126
126
126
133
136

$\Sigma XY = 1110$

$$\frac{(\Sigma X)(\Sigma Y)}{N} = \frac{(54)(189)}{9} = 1134$$

$$\Sigma xy = 1110 - 1134 = -24$$

5.
$$r = \frac{\Sigma xy}{\sqrt{(\Sigma x^2)(\Sigma y^2)}}$$

$$= \frac{-24}{\sqrt{(12)(48)}}$$

$$= \frac{-24}{24} = -1.00$$

6.

X	X^2	Y	Y^2	XY
236	55696	19	361	4484
304	92416	20	400	6080
97	9409	32	1024	3104
231	53361	24	576	5544
140	19600	27	729	3780
85	7225	35	1225	2975
168	28224	22	484	3696
107	11449	25	625	2675
171	29241	20	400	3420
250	62500	21	441	5250
171	29241	23	529	3933
250	62500	21	441	5250
231	53361	24	576	5544
260	50625	18	324	4680
97	53361	26	676	2522
156	67600	18	324	2808
240	9409	23	529	5520
111	24336	28	784	3108
225	57600	20	400	4500
231	12321	24	576	5544
$\Sigma X = 3761$	$\Sigma X^2 = 789,475$	$\Sigma Y = 470$	$\Sigma Y^2 = 11,424$	$\Sigma XY = 84,417$

A. $\Sigma xy = 84,417 - \dfrac{(3761)(470)}{20}$

$= 84,417 - 88,383.5$

$= -3966.5$

B. $\Sigma x^2 = 789,475 - \dfrac{(3761)^2}{20}$

$= 789,475 - 707,256.05$

$= 82,218.95$

C. $\Sigma y^2 = 11{,}424 - \dfrac{(470)^2}{20}$

$\quad = 11{,}424 - 11{,}045$

$\quad = 379$

D. $r = \dfrac{-3966.5}{\sqrt{(88{,}218.95)(379)}}$

$\quad = \dfrac{-3966.5}{5582.2} = -0.71$

7.

Score	Rank
120	3
142	2
110	4
90	5
67	6
185	1
54	7

8.

Score	Rank
15	1
5	5
9	4
10	3
11	2

9.

Score	Rank
87	1
72	2
64	3
53	4
19	5
5	6
2	7
1	8

10.

Score	Rank
115	3
86	8
119	2
73	10
136	1
62	11
105	5
112	4
75	9
97	7
104	6

11.

Rank on X	Rank on Y	D	D^2
1	8	−7	49
3	7	−4	16
2	6	−4	16
4	5	−1	1
7	4	+3	9
6	3	+3	9
5	2	+3	9
8	1	+7	49
		$\Sigma D = 0$	$\Sigma D^2 = 158$

$r_{\text{rho}} = 1 - \dfrac{6(158)}{8(63)}$

$\quad = 1 - \dfrac{948}{504}$

$\quad = -0.88$

12.

Rank on X	Rank on Y	D	D^2
1	1	0	0
2	2	0	0
3	3	0	0
4	4	0	0
5	5	0	0
		$\Sigma D = 0$	$\Sigma D^2 = 0$

$r_{\text{rho}} = 1 - \dfrac{6(0)}{5(24)}$

$\quad = 1 - 0$

$\quad = 1.00$

13.

Rank on X	Rank on Y	D	D^2
1	3	−2	4
2	1	1	1
3	6	−3	9
4	5	−1	1
5	2	3	9
6	4	2	4
		$\Sigma D = 0$	$\Sigma D^2 = 28$

$r_{\text{rho}} = 1 - \dfrac{6(28)}{6(35)}$

$\quad = 1 - 0.80$

$\quad = 0.20$

14.

Rank on X	Rank on Y	D	D^2
12	1	11	121
8	2	6	36
10	3	7	49
9	4	5	25
4	5	−1	1
6	6	0	0
1	7	−6	36
2	8	−6	36
7	9	−2	4
11	10	1	1
3	11	−8	64
5	12	−7	47
		$\Sigma D = 0$	$\Sigma D^2 = 422$

$$r_{rho} = 1 - \frac{6(422)}{12(143)}$$

$$= 1 - 1.48$$

$$= 0.42$$

15.

Score	Rank
18	2
12	6
10	8
11	7
14	4
19	1
14	4
4	9.5
4	9.5
14	4

16.

Score	Rank
108	3.5
101	6
115	1.5
108	3.5
115	1.5
103	5

17.

Rank on X	X^2	Rank on Y	Y^2	XY
1	1	1	1	1
3	9	2	4	6
3	9	3	9	9
3	9	4	16	12
5	25	5	25	25
$\Sigma X = 15$	$\Sigma X^2 = 53$	$\Sigma Y = 15$	$\Sigma Y^2 = 55$	$\Sigma XY = 53$

A. $\Sigma xy = 53 - \frac{(15)(15)}{5}$

$= 53 - 45$

$= 8$

B. $\Sigma x^2 = 53 - 45$

$= 8$

C. $\Sigma y^2 = 55 - 45$

$= 10$

D. $r = \frac{8}{\sqrt{(8)(10)}}$

$= \frac{8}{8.94}$

$= 0.89$

18.

Rank on X	X^2	Rank on Y	Y^2	XY
1	1	1.5	2.25	1.5
2	4	1.5	2.25	3.0
3.5	12.25	3	9	10.5
3.5	12.25	6	36	21.0
5	25	5	25	25.0
6.5	42.25	4	16	26.0
6.5	42.25	7	49	45.5
$\Sigma X = 28.0$	$\Sigma X^2 = 139$	$\Sigma Y = 28.0$	$\Sigma Y^2 = 139.5$	$\Sigma XY = 132.5$

A. $\Sigma xy = 132.5 - \dfrac{(28)(28)}{7}$

$= 132.5 - 112$

$= 20.5$

B. $\Sigma x^2 = 139 - \dfrac{(28)^2}{7}$

$= 139 - 112$

$= 27$

C. $\Sigma y^2 = 139.5 - \dfrac{(28)^2}{7}$

$= 139.5 - 112$

$= 27.5$

D. $r = \dfrac{20.5}{\sqrt{(27)(27.5)}}$

$= \dfrac{20.5}{27.25}$

$= 0.75$

CHAPTER 9 TEST

1. a
2. b
3. c
4. d
5. c
6. a
7. b
8. d
9. b
10. c
11. d
12. c
13. d
14. b
15. a
16. $r = 0.76$

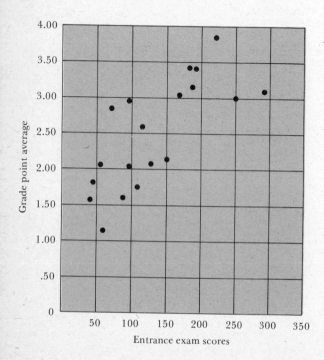

17.

Salesperson	Rank in assertiveness	Sales in thousands	D	D^2
A	1	1	0	0
B	2	6	-4	16
C	3	3	0	0
D	4	2	2	4
E	5	7	-2	4
F	6	4	2	4
G	7	8	-1	1
H	8	5	3	9
I	9	10	-1	1
J	10	12	-2	4
K	11	9	2	4
L	12	13	-1	1
M	13	11	2	4
				$\Sigma D^2 = 52$

$$r_{\text{rho}} = 1 - \frac{6(52)}{13(168)}$$

$$= 1 - \frac{312}{2184}$$

$$= 0.86$$

CHAPTER 10 EXERCISES

1. $Y' = 23$ 　　2. $Y' = 2.7$ 　　3. $Y' = 12.5$ 　　4. $Y' = 8.3$ 　　5. $Y' = 2$

6.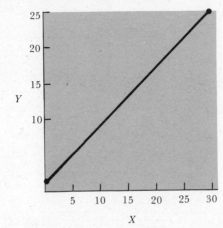

7. $b_{yx} = \frac{20}{15} = 1.33$ 　　8. $b_{yx} = \frac{-10}{10} = -1.00$

9. $b_{yx} = \frac{28}{59} = 0.47$ 　　10. $b_{yx} = \frac{0}{30} = 0.00$

11. $a_{yx} = 26 - 1.33(5)$ $= 13.35$ 　　12. $a_{yx} = 19.4 - (-1.00)(4.5)$ $= 23.9$

13. $a_{yx} = 142 - 0.47(8.7)$ $= 146.09$ 　　14. $a_{yx} = 101.3 - 0.00(65.4)$ $= 101.3$

15. $Y' = 13.35 + 1.33\,X$ 　　16. $Y' = 23.90 + (-1.00)\,X$

17. $Y' = 137.91 + 0.47 X$

18. $Y' = 65.40$

19. a) $Y' = 13.35 + 13.30$
      $= 26.65$
   b) $Y' = 13.35 + 53.20$
      $= 66.55$

20. a) $Y' = 23.90 - 20$
      $= 3.90$
   b) $Y' = 23.90 - 5$
      $= 18.90$

21. a) $Y' = 66.31 + 37.60$
      $= 103.91$
   b) $Y' = 66.31 + 65.80$
      $= 132.11$

22. a) $Y' = 65.40$
   b) $Y' = 65.40$

23.

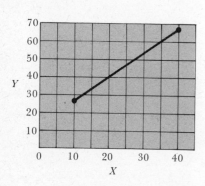

24.

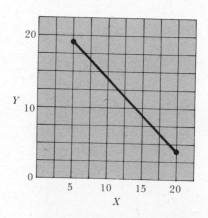

25.

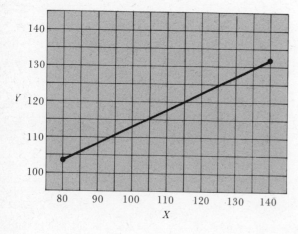

26.

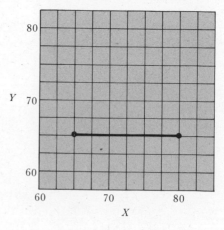

27. a) $X' = 16 + (-0.7)(25)$
      $= -1.5$
   b) $X' = 16 + (-0.7)(42)$
      $= -13.4$

28. $b_{xy} = \dfrac{20}{74} = 0.27$

29. $b_{xy} = \dfrac{-10}{15.62} = -0.64$

30. $b_{xy} = \dfrac{28}{82.35} = 0.34$

31. $a_{xy} = 5 - (0.27)(20)$
     $= -0.4$

32. $a_{xy} = 4.5 - (-0.64)(19.4)$
     $= 16.92$

33. $a_{xy} = 8.7 - 0.34(142)$
     $= -39.58$

34. $X' = -0.4 + 0.27Y$

35. $X' = 16.92 + (-0.64)Y$

36. $X' = -39.58 + 0.34Y$

37. a) $X' = -0.4 + 3.24$
      $= 2.84$
    b) $X' = -0.4 + 7.56$
      $= 7.16$

38. a) $X' = 16.92 - 5.12$
      $= 11.80$
    b) $X' = 16.92 - 19.20$
      $= -2.28$

39. a) $X' = -39.58 + 34$
      $= -5.58$
    b) $X' = -39.58 + 61.20$
      $= 21.62$

40.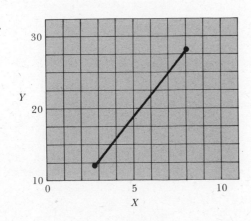

41.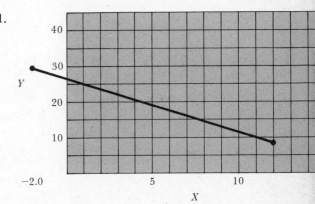

42. $X' = 188.05 + 0.85 \dfrac{65.78}{6.05} (34.49 - 26.65)$

    $= 188.05 + 72.46$
    $= 260.51$

43. $s_{\text{est } X} = 0.00$

44. $s_{\text{est } X} = 6.67$

45. $s_{\text{est } X} = 9.75$

46. $s_{\text{est } X} = 11.11$

47. $s_{\text{est } X} = 12.57$

48. $s_{\text{est } X} = 12.83$

49. $s_{\text{est } X} = 12.69$

50. $s_{\text{est } X} = 12.24$

51. $s_{\text{est } X} = 9.16$

52. $s_{\text{est } X} = 4.01$

53. $s_{\text{est } X} = 0.00$

54. $s_{\text{est } Y} = 0.29$

55. $s_{\text{est } Y} = 3.72$

56. $s_{\text{est } Y} = 19.80$

57. $s_{\text{est } Y} = 0.29$

58. $z_{xy} = \dfrac{115 - 105}{6} = 1.67$

59. $z_{xy} = \dfrac{80 - 97}{8} = -2.12$

60. $z_{xy} = \dfrac{103 - 100}{12} = 0.25$

61. Percentile rank = 10.93

62. Percentile rank = 95.05

63. Percentile rank = 26.76

64. $z_{yx} = \dfrac{42 - 60}{21.40} = -0.84$

65. $z_{yx} = \dfrac{138 - 100}{16.42} = 2.31$

66. $z_{xy} = \dfrac{8.63 - 8.93}{2.46} = -0.12$

67. Percentile rank = 20.05

68. Percentile rank = 98.96

69. Percentile rank = 4.78

70. A. $Y' = 1.56 + 0.70 (0.80)(115 - 100)$
       $= 1.56 + 0.56$
       $= 2.12$

    B. $s_{\text{est } Y} = 0.57$

    C. $z_{yx} = \dfrac{3.51 - 2.12}{0.57} = 2.44$

Margaret O's percentile rank, among those for whom $Y' = 2.12$, is 99.27. Her performance is excellent.

CHAPTER 10 TEST

1. c	2. a	3. d	4. c	5. c
6. b	7. b	8. c	9. b	10. a
11. a	12. b	13. a	14. a	15. d

16. When $Y = 20$, $X' = 3.00$; when $Y = 80$, $X' = 8.84$.

When $X = 2.92$, $Y' = 37$; when $X = 8.92$, $Y' = 63$.

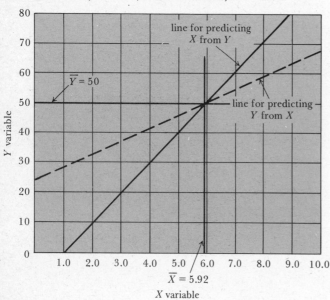

17. a) $b_{yx} = -1.20$ b) $a_{yx} = 49.08$

$b_{xy} = -0.56$ $a_{xy} = 33.68$

18.	$s_{\text{est } X}$	$s_{\text{est } Y}$
a)	17.84	1.46
b)	7.55	12.99
c)	91.45	26.07
d)	8.39	3.93

19. $Y' = 2.88 + 0.60 \left(\dfrac{0.79}{100} \right) (480 - 500)$

$= 2.79$

$s_{\text{est } Y} = 0.63$

$z = \dfrac{3.95 - 2.79}{0.63} = 1.84$

G. John is peforming far better than expected. His percentile rank is 96.71 among those who entered his school with a college entrance of 480.